Small Wars and Insurgencies in Theory and Practice, 1500–1850

In early modern times, warfare in Europe took on many diverse and overlapping forms. Our modern notions of 'regular' and 'irregular' warfare, of 'major war' and 'small war', have their roots in much greater diversity than such binary notions allow for. While insurgencies go back to time immemorial, they have become conceptually fused with 'small wars'. This is a term first used to denote special operations, often carried out by military companies formed from special ethnic groups and then recruited into larger armies. In its Spanish form, *guerrilla*, the term 'small war' came to stand for an ideologically-motivated insurgency against the state authorities or occupying forces of another power.

There is much overlap between the phenomena of irregular warfare in the sense of special operations alongside regular operations, and irregular warfare of insurgents against the regular forces of a state. This book demonstrates how long the two phenomena were in flux and fed on each other, from the raiding operations of the sixteenth century to the 'small wars' or special operations conducted by special units in the nineteenth century, which existed alongside and could merge with a popular insurgency.

This book is based on a special issue of the journal *Small Wars & Insurgencies*.

Beatrice Heuser holds the chair in International Relations at the University of Reading, UK. Her research focuses on strategy, European security, transatlantic relations, Britain, France, USA, Germany, and defence policy making. She has published on nuclear strategy, Clausewitz, and the evolution of strategy since Antiquity.

Small Wars and Insurgencies in Theory and Practice, 1500–1850

Edited by
Beatrice Heuser

LONDON AND NEW YORK

First published 2016 by Routledge

2 Park Square, Milton Park, Abingdon, Oxfordshire OX14 4RN
711 Third Avenue, New York, NY 10017

Routledge is an imprint of the Taylor & Francis Group, an informa business

First issued in paperback 2017

British Library Cataloguing in Publication Data
A catalogue record for this book is available from the British Library

ISBN 13: 978-1-138-94167-0 (hbk)
ISBN 13: 978-1-138-29978-8 (pbk)

Typeset in Times
by RefineCatch Limited, Bungay, Suffolk

Publisher's Note
The publisher accepts responsibility for any inconsistencies that may have
arisen during the conversion of this book from journal articles to book chapters,
namely the possible inclusion of journal terminology.

Disclaimer
Every effort has been made to contact copyright holders for their permission to
reprint material in this book. The publishers would be grateful to hear from any
copyright holder who is not here acknowledged and will undertake to rectify
any errors or omissions in future editions of this book.

Contents

CONTENTS

Citation Information

The following chapters were originally published in *Small Wars & Insurgencies*, volume 25, issue 4 (August 2014). When citing this material, please use the original page numbering for each article, as follows:

Preface
'The Origins of Small Wars from Special Operations to Ideological Insurgencies': A National Army Museum response
Alastair Massie
Small Wars & Insurgencies, volume 25, issue 4 (August 2014) pp. 738–740

Introduction
Exploring the jungle of terminology
Beatrice Heuser
Small Wars & Insurgencies, volume 25, issue 4 (August 2014) pp. 741–753

Chapter 1
The sixteenth-century antecedents of special operations 'small war'
Benjamin Deruelle
Small Wars & Insurgencies, volume 25, issue 4 (August 2014) pp. 754–766

Chapter 2
The essence of war: French armies and small war in the Low Countries (1672–1697)
Bertrand Fonck and George Satterfield
Small Wars & Insurgencies, volume 25, issue 4 (August 2014) pp. 767–783

Chapter 3
Initiating insurgencies abroad: French plans to 'chouannise' Britain and Ireland, 1793–1798
Sylvie Kleinman
Small Wars & Insurgencies, volume 25, issue 4 (August 2014) pp. 784–799

Chapter 4
The insurgency of the Vendée
Alan Forrest
Small Wars & Insurgencies, volume 25, issue 4 (August 2014) pp. 800–813

Chapter 5
Guerrillas and bandits in the Serranía de Ronda, 1810–1812
Charles Esdaile
Small Wars & Insurgencies, volume 25, issue 4 (August 2014) pp. 814–827

Chapter 6
The German wars of liberation 1807–1815: The restrained insurgency
Martin Rink
Small Wars & Insurgencies, volume 25, issue 4 (August 2014) pp. 828–842

Chapter 7
Poachers turned gamekeepers: A study of the guerrilla phenomenon in Spain, 1808–1840
Mark Lawrence
Small Wars & Insurgencies, volume 25, issue 4 (August 2014) pp. 843–857

Chapter 10
Lessons learnt? Cultural transfer and revolutionary wars, 1775–1831
Beatrice Heuser
Small Wars & Insurgencies, volume 25, issue 4 (August 2014) pp. 858–876

The following chapter was originally published in *The Journal of Strategic Studies*, volume 33, issue 1 (February 2010). When citing this material, please use the original page numbering for each article, as follows:

Chapter 8
Small Wars in the Age of Clausewitz: The Watershed Between Partisan War and People's War
Beatrice Heuser
The Journal of Strategic Studies, volume 33, issue 1 (February 2010) pp. 139–162

The following chapter was originally published in *Civil Wars*, volume 14, issue 1 (March 2012). When citing this material, please use the original page numbering for each article, as follows:

Chapter 9

Atrocities in Theory and Practice: An Introduction
Beatrice Heuser
Civil Wars, volume 14, issue 1 (March 2012) pp. 2–28

For any permission-related enquiries please visit:
http://www.tandfonline.com/page/help/permissions

Notes on Contributors

Benjamin Deruelle is a lecturer, and a researcher at the Institut de Recherche Historique du Septentrion (IRHiS), at the University of Lille, France. He works on culture, representations, and practices of war in the first French modernity. He is about to publish his thesis, 'Of Paper, Iron and Blood: Knights and Knighthood confronted with the Sixteenth Century (ca. 1460 – ca. 1620)', with Publications de la Sorbonne, Paris, France. He is also the author of articles on the question of the revival of the chivalric ideal and its political, military, and social issues.

Charles Esdaile holds a personal chair in the Department of History at the University of Liverpool, UK. Britain's leading expert on the Peninsular War of 1808–1814, he is the author of numerous books on the Napoleonic period, including *The Spanish Army in the Peninsular War* (1988), *The Duke of Wellington and the Command of the Spanish Army, 1812–1814* (1990), *The Wars of Napoleon* (1995), *The Peninsular War: a New History* (2002), *Fighting Napoleon: Guerrillas, Bandits and Adventurers in Spain, 1808–1814* (2004), *Napoleon's Wars* (2007), *Peninsular Eyewitnesses: the Experience of War in Spain and Portugal, 1808–1814* (2008), and *Outpost of Empire: the French Occupation of Andalucía, 1810–1812* (2012).

Bertrand Fonck is a graduate from the École des chartes, and is a curator in the Service historique de la Défense, Vincennes, France, where the archives of the French Ministry of Defence are conserved, from the 17th century to the present. He holds a PhD in History from the University of Paris-Sorbonne. He is the author of *Le maréchal de Luxembourg et le commandement des armées sous Louis XIV* (Seyssel, Champ Vallon, 2014). He has co-edited several collective volumes including *La fin de la Nouvelle-France* (Paris, Armand Colin, 2013), *Combattre et gouverner. Dynamiques de l'histoire militaire de l'époque moderne* (Rennes, Presses universitaires de Rennes, 2015) and *L'âge d'or de la cavalerie* (Paris, Gallimard, 2015).

Alan Forrest is emeritus professor of Modern History at the University of York, UK. He has published widely on modern French history, especially on the French Revolution and Empire, and on the history of war. His works include *Napoleon's Men: The Soldiers of the Revolution and Empire* (2002), *Paris, the Provinces, and the French Revolution* (2004), *The Legacy of the French Revolutionary Wars: The Nation-in-Arms in French Republican Memory* (2009), *Napoleon* (2011), and *Waterloo* (2015). He is editor, with Etienne François and Karen Hagemann, of *War Memories: The Revolutionary and Napoleonic Wars in Modern European Culture* (2012), and, with Matthias Middell, of *The Routledge Companion to the French Revolution in World History* (forthcoming, 2016).

Beatrice Heuser was educated at the universities of London and Oxford, UK, where she obtained her DPhil in International Relations. She taught at the Department of War Studies, King's College London, UK (1991–2003), and since 2007, she has held the chair in International Relations at the University of Reading, UK. She has also been Director of Research at the Military History Research Office of the Bundeswehr in Potsdam, Germany (2003–2007), and worked for NATO's international staff (1996–1997). Her publications include *Nuclear Strategies and Forces for Europe* (1997), *Nuclear Mentalities? Strategies and Beliefs in Britain, France and the FRG* (1998), *The Bomb* (1999), *Reading Clausewitz* (2002), and *The Evolution of Strategy: Thinking War from Antiquity to the Present* (2010).

Sylvie Kleinman holds an MPhil and a PhD, and is a research associate at the Centre for War Studies, and an occasional lecturer in the Department of History at Trinity College, Dublin. Her areas of expertise include Ireland 1775–1847, and Slavery in History and Memory. From 2007 to 2009, she was an IRCHSS post-doctoral fellow at Trinity College Dublin. Her research, publications, and teaching cover military-civilian relations during occupations, eighteenth century martial and national cultures, and the life and legacy of Theobald Wolfe Tone (1763–1798).

Mark Lawrence is a lecturer in Military History (1700–1900) at the University of Kent and has previously worked at the universities of Newcastle-upon-Tyne, Sheffield, Liverpool, and the Autónoma de Zacatecas. He is a historian of civil war, radicalism, and society in modern Spain and Mexico and holds a PhD (Liverpool, 2008) on the origins of Spanish radicalism. His publications include *Spain's First Carlist War, 1833–40* (Palgrave, 2014). He is currently completing a comparative study of civil war in modern Spain and is also projecting a study of Cristero and Agrarian conflict in central-northern Mexico. He also has a professional background in literary translation and interpreting in German and Spanish.

Martin Rink served as an officer in the Bundeswehr 1985–1996. From 1989 to 1993 he studied History, Political and Social Sciences at the University of the Bundeswehr in Munich, then obtained a doctorate in History at the University of Potsdam in 1998. His published doctorate received a 2000 Werner-Hahlweg-Prize for Military History. He is a historian at the Centre for Military History and Social Sciences of the Bundeswehr, while teaching at the University of Potsdam and the University of the Bundeswehr in Munich. His publications include *Die Bundeswehr 1950/55–1989* (Munich, 2015), "The Service Staffs' Struggle over Structure: The Bundeswehr's internal Debates on adopting NATO Doctrine 1950–1963", in: James C. Corum (ed.), *Rearming Germany* (Amsterdam, 2011), p. 221–251, *Is There Anything New in Small-Scale Warfare? Developments in Asymmetric Violence, 1740–1815*, Working Papers in Military and International History, No. 6 (Salford/Manchester, 2009), "The Partisan's Metamorphosis from Freelance Military Entrepreneur to German Freedom Fighter, 1740–1815", in: *War in History*, No. 17, January 2010, p. 6–36, *Vom Partheygänger zum Partisanen. Die Konzeption des kleinen Krieges in Preußen 1740–1813* (Frankfurt am Main, 1999).

George Satterfield holds a PhD in History from the University of Illinois, Urbana-Champaign, IL, USA, and is presently an associate professor at the U.S. Naval War College, Newport, Rhode Island, USA. He is the author of *Princes, Posts, and Partisans: The Army of Louis XIV and Partisan Warfare in the Netherlands, 1673–1678* (2003). He has also contributed diverse articles on military history, including on the subjects of irregular warfare and revolutions in military affairs.

PREFACE

'The Origins of Small Wars from Special Operations to Ideological Insurgencies': A National Army Museum response

Alastair Massie

National Army Museum, London, UK

The National Army Museum in Chelsea, London was delighted, on 25 March 2013, to host the conference 'The Origins of Small Wars: From Special operations to Ideological Insurgencies'. It was the first time that the Museum had sponsored such an ambitious event, with speakers drawn from France, the United States, Germany, as well as the United Kingdom. The new readiness of the Museum to do so is indicative of the way in which the approach taken to its work has recently changed; a fresh approach which in turn stems from the successful application made by the Museum to the Heritage Lottery Fund (HLF) in May 2012 for an initial grant to undertake the complete redevelopment of the site in Chelsea. Assuming that the HLF approves the final plans – and matching funding is raised – the Museum will close in April 2014 for a two-year period to enable its interior to be removed and new galleries installed. The project will cost £22,750,000.

Clearly, prospective donors are unlikely to be attracted to the enterprise if the intention were simply to tell the story of the British Army in the same chronological fashion that the Museum currently does, or even as it has in the past, with galleries devoted to specific types of exhibit – paintings, weapons, uniforms, and medals. Consequently, an ambitious new interpretation strategy has been put in train, which will see the Museum's displays grouped in themes. These include 'Soldier', looking at the serviceman's life cycle (recruitment, training, daily life, combat, and demobilisation); 'Battle', its planning, preparation, execution, aftermath and the change and innovation that results from the British Army's key engagements; 'Army', what the British Army is, how it is organised, what it is like to be part of, and how it has changed over time; 'Society', examining the Army's relationship with civilians and the responses provoked; and 'History', intended to guide the visitor in discovering more not only about the British Army and its impact on national and world events, but how to ask the right questions when it comes to finding out about the Army and their family's past.

A thematic approach to telling the story of the British Army is a tricky thing to carry off: the certainties of a linear treatment of events are no longer available. If it is to be done convincingly, it has to be thought through and a premium placed on expert knowledge. This explains the National Army Museum's new-found interest in academic conferences. What better way to access the latest and best thinking on a given topic? Indeed, to better facilitate this engagement with the outside world, the entire organisational structure of the Museum was reconfigured in 2012, the old curatorial departments broken up and replaced by departments of access and outreach (among others). I myself ceased to be Head of Archives and Photographs and became Head of Academic Access instead. As such, I have been given the remit to organise conferences and, following on from 'Origins of Small Wars', two further conferences on the cultural afterlife of the Crimean War and 'Making Military History in Museums' are scheduled for 2013.

What then did we at the National Army Museum learn from 'Origins of Small Wars'? Bearing in mind some of the themes identified above, the work of Hervé Drévillon on territorial control in France is relevant to the gallery provisionally designated 'Society', inviting comparisons with the role of the British Army in Ireland during the same era. The papers of George Satterfield and Bertrand Fonck – the former showing how partisans operating in the Spanish Netherlands in the 1670s were increasingly valued as an adjunct to French regular forces, and the latter detailing how Marshal Luxembourg eventually clasped them to his bosom – are interesting illustrations of the ways in which armies change (the theme of our intended third gallery). If one were to draw comparisons with the British experience, it would date from a century later, when the potential of the irregulars who had served in North America in the 1750s was only fully realised by Sir John Moore and his Shorncliffe-trained light infantry over 40 years afterwards.

I was struck too by the dehumanising concept of one's opponents as 'other' in many of the small wars of the period. This came across particularly strongly in Alan Forrest's paper on the insurgency in the Vendée, was touched upon by Charles Esdaile in his analysis of events in Andalusia during the French occupation, and was implicit when Christopher Duffy spoke about the Jacobite Rebellion of 1745. Because the British Army enjoyed the widest experience of extra-European campaigning, it had perhaps the greatest opportunity to dehumanise the enemy, something given its fullest rein during the suppression of the Indian Mutiny in 1857–1858. These, however, are the kinds of difficult issues which the National Army Museum is nowadays expected to treat in its displays and it was instructive to hear them aired, in a comparative context, during the conference. In the same way, it was illuminating to hear both Charles Esdaile and Martin Rink talk about national mythologising in relation to small wars, explaining how what occurred in the Serrania de Ronda in 1810–1812 was not an expression of guerrilla resistance but simply a reaction to French foraging, and how the 1813 war of liberation in Germany was, ultimately, but an imagined insurgency. This is useful to us because it will inevitably fall to the Museum to perform a similar exercise in its representation of the Indian Mutiny, tactfully pointing out the ahistoricism of terming it the 'First War of Indian Independence' when the description 'Mutiny of the Bengal Army' is more accurate.

The National Army Museum is highly gratified that the excellence of the papers presented during the conference will leave a permanent mark in the pages of *Small Wars and Insurgencies*. Our thanks are due to Professor Beatrice Heuser not only for facilitating this, but for suggesting the conference's theme in the first place and enlisting the majority of the speakers.

Introduction: Exploring the jungle of terminology

Beatrice Heuser

Department of Politics & IR, University of Reading, Reading, UK

When twentieth-century authors wrote about 'partisan warfare', they usually meant an insurgency or asymmetric military operations conducted against a superior force by small bands of ideologically driven irregular fighters. By contrast, originally (i.e. before the French Revolution) 'partisan' in French, English, and German referred only to the leader of a detachment of special forces (party, *partie*, *Parthey*, *détachement*) which the major European powers used to conduct special operations alongside their regular forces. Such special operations were the classic definition of 'small war' (*petite guerre*) in the late seventeenth and in the eighteenth centuries. The Spanish word *'la guerrilla'*, meaning nothing other than 'small war', only acquired an association with rebellion with the Spanish War of Independence against Napoleon. Even after this, however, armies throughout the world have continued to employ special forces. In the late nineteenth century, their operations have still been referred to as prosecuting *'la guerrilla'* or 'small war', which existed side by side with, and was often mixed with, 'people's war' or popular uprisings against hated regimes.

The present journal has done more than any other publication of the recent past to define the parameters of 'small wars', 'insurgencies', and counter-insurgency. This has been very valuable, not least because there are few areas of strategic studies where the semantics are so complex and the terminology used so diverse, conflicting, overlapping, or else vague and confusing. Much ink has been spilled on what constitutes asymmetric warfare, low-intensity conflict, operations other than war, or the differences between counter-insurgency and counter-guerrilla. 'Exterior operations' or 'operations abroad' are at best bureaucratically convenient catch-all terms. These are at best 'new wars' in small technological aspects such as the use of mobile phones or cyberspace. It is to the credit of this journal that its editors have plumped for the classical terms 'small wars and insurgencies'. But those, too, lend themselves to multiple interpretations. Some can best be explained by looking at the historical evolution of the use of these terms. Their origins are largely forgotten, however, and their early use is often ignored. This special issue is dedicated to shedding light on the origins and diversity of these terms in their early usage until the early nineteenth century.

Partisan units as special forces

Throughout the recorded history of warfare, we find instances where writers contrast 'regular' fighting, warfare following rules, often large-scale warfare, with 'irregular' warfare, where rules are abandoned, often practised by smaller units against numerically superior ones. Sometimes the latter form of warfare was used to complement the former; often, in the absence of viable alternatives, it was used by weaker parties against stronger forces. By the time the Romans clashed with the Carthaginians, Hannibal had identified the distinctiveness in fighting style and the particular usefulness of the Nubian cavalry, and he used it for swift surprise operations on the margins of campaigns to complement his regular forces designed to meet adversaries in set-piece battles. Rome followed suit and began to use *socii* (allies) and other auxiliary forces according to their special skills. The East Romans equally ruled diverse populations with particular styles in warfare. Endowed with this meaning, the Byzantine term *stratiotoi* – originally the peasant who was given land in return for military service in the late tenth century – lived on after the fall of the East Roman Empire in 1465. The north Italian states called their Albanian light infantry who fought alongside other mercenaries '*stradioti*'.[1]

Not surprisingly, the Roman-*cum*-Byzantine pattern of using the special talents of the different ethnic groups under imperial rule was replicated by the Holy Roman, Russian, and Ottoman Empires, which saw themselves as the successor to Byzantium, each with its many subject peoples. In 1688, the Habsburgs set up their first regiments of special Hungarian light cavalry, the original Hussars. In the following century, Habsburg empress Maria Theresa also recruited Croats, who were regarded as particularly good at clandestine operations,[2] as well as 'Sclavonians, Waradians, Licanians, ... Rascians, Banatistes & Pandours', peoples and tribes washed up in the Balkans during the great migrations, some of longer local lineage, most of whom have by now all but disappeared in cauldrons of twentieth-century enforced national homogenisation. Serbs, Bosniaks, and Vlachs were also famed for their special martial skills, particularly for warfare in mountainous areas, while the Cossacks, fighting much in the tradition of the equestrian peoples of the great steppes, as well as Tatars, Pandures, and Turkic tribes were recruited by the Russian tsars for special operations.

Other European monarchs emulated these examples. In 1744, the French, imitating the Habsburgs, raised several regiments of special forces and *compagnies franches* (free companies), or special *partis* (partisan corps), to fight as *troupes légères* (light troops or forces), mostly on horseback, including French Hussars, in the same way as those South-East European partisan forces of Her Imperial Majesty.[3]

In Western Europe, Scots and Swiss had also long served foreign polities as special forces. Already in the fifteenth century, Scots had served the French crown in special fighting units. In 1419, during the Hundred Years War, the Scot John Stuart of Darnley and a Scottish contingent were sent by the Scottish king to fight for Charles VII of Valois in his struggle against the Plantagenet Henry V. Along with a Count Douglas and a Count Buchan, Darnley was rewarded with French lands and titles, and made Count of Evreux. His Scottish soldiers became the first *gens d'armes* unit of King Charles VII in 1422, as perhaps the first French standing forces.[4] The Hundred Years War drew in mercenaries from many other areas as well, and the Swiss mercenaries owed their professional reputation not least to it. They fought as hirelings for Lorraine against the Burgundians in the battle of Nancy (1477) in which Charles the Bold of Burgundy was killed. When the king of Hungary, Matthias Corvinus, set up and paid for a standing army (the 'Black Army') in the second half of the fifteenth century, Swiss mercenaries were drawn on for their expertise. They were often found wherever polities tried to fend off the expansion of the Habsburgs, e.g. in Northern Italy. Most famously, to this day, they serve as guards to the small polity of the Vatican. Recruits from the German-speaking lands of the Holy Roman Empire, known as *Landsknechte*, were also frequently employed throughout Europe. They were feared for their indiscipline and excesses. All of these excelled as infantrymen, e.g. Scots and Swiss fighting with pikes in tight formations called *schiltrons* or *Haufen*. Some *Landsknecht* formations specialised in cutting down the pikes of their opponents with swords requiring two hands and great body strength to wield – the *Zweihänder* – and became known as *verlorner Haufen* or *verlorne Rotte* (the lost unit), in French *enfants perdus*, as this was largely a suicide mission.

By contrast, special forces in central and eastern Europe usually consisted of light cavalry or a mix or light cavalry and loosely deployed infantry. The tasks of such light cavalry and infantry units tended to be roughly threefold: gathering intelligence, securing provisions, and dirty tricks. The French Captain Grandmaison in his great classic on *Small War, or Treatise on the Service of Light Troops on Campaigns* of 1756 explained that it was useful for France also to have units of foreigners among its '*compagnies franches*' (another term Grandmaison employed for '*partis*' or '*détachements*'[5]), especially those that hailed from the areas where France fought her wars – Germans, Flemings, Italians – as they could serve as guides, interpreters, scouts, and especially spies.[6] Intelligence of all sorts was key: Where were the enemy forces? What were they about to do (taking which route, with possible places along the route lending themselves to an ambush)? Where were they camped or billeted? Did they have many sentries? Were villages and towns well guarded (especially at night)? At what time did the guards change? What were the defences like? Were reinforcements or convoys with food or other supplies expected, and if so, when and coming from where? Counter-intelligence was also important: the leader of the *parti*, the *Partisan*, should keep a watchful eye on spies hailing from the theatre of operations, and guard against being fed false information.[7]

Securing victuals for the soldiers and fodder for the horses was also vital, usually by way of forced contributions from locals in money or goods. *Compagnies franches* were to bring in these provisions by force or ruse: disguised as peasants, they might lead enemy forces into an ambush with the hope of loot or of snatching the enemy's food and fodder.[8]

The harassment of the adversary was also important in itself, through inflicting pinpricks on his regular forces, attacking rearguards and stragglers as much as advance parties. The recommended time for ambushes was during the night, the favourite place woods or geographic formations which lent themselves to these operations.[9] If it came down to raids on villages where enemy troops were quartered, Grandmaison had no qualms about recommending shooting anything that moved in such a hapless village, and setting fire to the houses.[10] (Martin Rink has rightly concluded that, before 1775, European partisan forces did not care much about the well-being of the civilian populations in their areas of operations.[11])

Grandmaison was probably the most widely read author on partisan warfare, but he was not the first (before him, there were the works of Antoine de la Ville and one De La Croix[12]), nor by far the only one. The earliest formal literature describing this sort of warfare can be traced to the seventeenth century.[13] A German-language encyclopaedia of 1740, i.e. before the publication of Grandmaison's work, noted that detachments ('*Parthey, Parti*') might be created to 'harass the enemy through ruse and speed, or to gather intelligence about him'.[14] In general usage, the term 'partisan' referred only to the leader of these special light forces.[15] The sizes of such detachments varied greatly, depending on the mission, but they were rarely smaller than 100 or larger than 2000 men.

As Benjamin Deruelle's contribution in this issue shows, these tasks and practices existed long before the terms '*petite guerre*'/'*kleiner Krieg*'/'*guerrilla*'/ 'small war', 'partisan' and '*parti(e)*'/'party'/'*Parthey*'/'detachments', or '*compagnie franche*' were commonly associated with them. In the sixteenth century, armies were much smaller overall than in the golden age of partisan units as special forces (the late seventeenth and eighteenth centuries). Consequently, armies had less of a capacity to designate units for special tasks only. Benjamin Deruelle demonstrates that for France at any rate, in the sixteenth and early seventeenth centuries, all soldiers could in principle be called upon to serve in a variety of functions, including those which later would become primarily the domain of specialised forces. These included foraging, gathering intelligence, and harassing the enemy while eschewing major encounters.

The pioneering works of Martin Rink and Sandrine Picaud-Monnerat have demonstrated that by the late seventeenth and eighteenth centuries, as armies had grown from five-figure to six-figure strength, the way of fighting of such detachments was generally distinct from that of regular forces and no longer contained task sets with which all forces were confronted sooner or later. For example, while the regular soldiers of the eighteenth century would be drilled to follow a very intricate routine of movements to enable them to fire successive salvos in short order, with all the men in the first and second lines firing, and those in the back lines reloading their muskets, the members of 'parties' would fire well-aimed single shots, in what was called *tirailleur*-tactics. They avoided all open terrain, preferring marshes, mountainous areas, or forests.[16] Today we would call them 'special forces'. They were used by states both for external and internal military operations, in the latter sense becoming the ancestors of the French gendarmerie.

Small war was thus, for the authors of the eighteenth and early nineteenth centuries, always part of a large-scale war, of conventional (symmetrical) operations elsewhere. The Prussian General Georg Wilhelm Freiherr von Valentini (1775–1834) wrote:

> What I understand by the so-called small war are all those actions in war, which merely support the operations of an army or a corps, without having an immediate effect on the conquest or defence of territory: the securing or even hiding of the main fighting force, both when it is stationary and when it is on the move, and those skirmishes which merely aim to harass the enemy.
>
> While it may seem as though the results of the small war have no substantial effect on the outcome of the war, they are still important to create the necessary conditions for affecting the final purpose of the war, the purpose of conquering a country or defending it.
>
> As a successfully conducted small war weakens the strength of the enemy and robs him of his means needed to prevail against us in the field [of battle], [a small war] can even decide the final result... [17]

Nor did the employment of special mercenary units, often but not exclusively recruited from one region outside the state that employs them, cease with the Age of Nationalism, ushered in by the French Revolution. The British East India Company employed Gurkha mercenaries at the latest by 1817. Gurkha hirelings fought to defend British interests in the subcontinent in the Anglo-Sikh Wars. As the British government took on the defence obligations of the Company in subsequent decades, it extended this pattern of hire, which has lasted until this day. The French Foreign Legion was established 1831, recruiting particularly from inhabitants of France's North African possessions, while for its home defence, successive French Republics had embraced the Roman ideal of the native citizen-soldier that had been endorsed through the ages by Machiavelli, Guibert, and the strategists of the French Revolution. Again, the Foreign Legion continues to be an instrument of the French state to this day, designated to fulfil special missions.[18] Many other states have in more recent times employed foreign professional soldiers, now known as Private Military Companies. In parallel, special forces recruited mainly within the state employing them have existed without interruption. To give just two examples of the latter, the British Long Range Desert Group and the British Special Air Services (better known by their acronym SAS) were set up to fight German forces in North Africa during the Second World War.[19] After being briefly broken up after 1945 the SAS was reformed in the early 1950s.

In general, then, small war, or partisan war, even in the twentieth and twenty-first centuries, was far more often than not part of a bigger war, a symmetrical encounter between 'regular' armed forces elsewhere. The decisive victories in all wars with a 'small war' component were not won by the partisans, but on battlefields, as George Satterfield and Bertrand Fonck demonstrate with their article in this issue. So much for the paradigm of 'small war' as special operations. Let us move now to the second paradigm, which emerged during the period to which this special issue is devoted: 'small war' as 'people's war'.

'*La Guerrilla*': 'small war' comes to mean 'insurgency'

Endless semantic confusion arises from the fact that the Spanish expression for 'small war', *la guerrilla*, used for the Spanish resistance against Napoleon's occupying forces in the Peninsular War, became associated also with something that was, in terms of motivation and incentive, if not necessarily in tactics, quite distinct from such 'special operations'. Moreover, whereas the term 'partisan' had previously designated only the leader of a *parti* or detachment, during this war it came to denote a group of irregular soldiers. The terms 'partisans' or '*guerrilla* fighters' have since then been used to refer to fighters recruited from among a population that refused to accept a foreign military occupation or a domestic regime in control of the (regular) military and the police.

The mobilisation of the population for warfare in itself was nothing new, and preceded the creation of professional armies. Throughout recorded history, the recruitment of fighters took place on a sliding scale between (or in a mix of) two basic models: on one end, the mobilisation of all able-bodied (mainly male) members of a society to fight for it (or for their lords); on the other end, the mercenary, the paid professional soldier, who may or may not hail from the social entity he is paid to defend. The *anciens régimes* of the eighteenth century to some degree mixed the two. The armies of Frederick II ('the Great') of Prussia consisted both of conscript soldiers, essentially lads drawn from the rural areas of the Hohenzollern possessions, and of hirelings from various parts of Europe, including non-German speaking parts of Europe like France. In marginal parts of Europe, border areas frequently fought over, or in European dependencies overseas, a militia could be organised to complement any standing army, and this militia would be mobilised in times of crisis. In the first half of the eighteenth century, the Jacobite rebellions against the Hanoverian kings of Scotland and England relied on locally levied peasants and shepherds who had been mobilised through a medieval feudal system of military service owed to the lord/laird, in Scotland still expressed through the tribal clan system.[20] Local farmers had been organised or had organised themselves for the defence of their land in the eighteenth century wars in North America, complementing the regular forces on both sides. In the American War of Independence against Britain (1775–1783), the militia element on the rebel side became particularly prominent.

The phenomenon of such an armed popular uprising presumably existed in all cultures since the earliest wars, and logically precedes the division between combatants and non-combatants. It was famously resuscitated by the French Revolution in 1793, with the *levée en masse*, as Alan Forrest has shown in his works.[21] Already Guibert, the great French strategist of the Enlightenment, had called for this sort of mobilisation of citizen-soldiers against a foreign threat, and he has consequently been seen as prophet of the French Revolution.[22] Its untrained volunteers could not meet regular soldiers with 'regular' tactics, as they lacked the latter's long training. Instead, they resorted quite naturally to the harassment tactics of the 'small war', sniping, laying ambushes, and attacking small units, rather than seeking a pitched battle. The Austrian Archduke Charles thus described the essence of these wars as not being about 'confronting the enemy's operations head-on, but merely to upset them through diversions'.[23] And yet some battles ensued in which the French Revolutionary forces encountered and prevailed over the regular forces of their adversaries, most famously perhaps the Battle of Valmy (20 September 1792) which was said to herald a 'new era' (in the words of Goethe[24]).

At the same time, a pattern of fighting comparable to that of the popular mobilisation of the American War of Independence and of the French Revolution can be found also in *resistance* movements *against* the French Revolutionaries. The first of these is the anti-revolutionary insurrection in the French area of Vendée, which began in 1793, which is examined in detail in Alan Forrest's contribution to this special issue, and the connected 'Chouannerie' uprising in Britanny.[25] Napoleon's campaigns engendered further reactions throughout Europe: they served as catalysts for the creation of a xenophobic nationalism which became the new dominant ideology for the Western world, whether in coexistence with old loyalties to king and country, with liberal views of human enlightenment and emancipation, or with republican rejection of any monarchy. Interestingly, there was also the opposite phenomenon, of France trying to export its revolution, by trying to incite a popular uprising in the United Kingdom's western possessions – mainly Ireland and Wales – creating a 'Chouannerie' directed against the rule of London. In some ways this was quite similar to French support given to rebels against the British monarchy during the Jacobite rebellions, most recently in 1745–1746, when the French crown paid Irish mercenaries, the Wild Geese, to join the rebels under the Young Pretender, Charles Edward Stuart. Supporting rebels within other polities, indeed inciting rebellions, thus giving such domestic struggles an interstate dimension, can be traced back to Antiquity, and was a prominent feature of the Wars of Religion from the Hussite Wars (1419–c.1434) to the Jacobite rebellions (1689–1746, which had an emphatically religious dimension). The attempts of the French Revolutionary government to 'chouannise' Britain and Ireland, however, took on a new ideological dimension in rhetoric if not in intention distinguishing it from the previous religious wars, as Sylvie Kleinman's contribution to this special issue shows. They were thus arguably the first example of an ideological support by one state for rebels in other states, a pattern that would gain considerable importance in the twentieth century, and thus quite a turning point in the history of small wars and insurgencies.

Turning back to the *anti*-French insurgencies, the most famous anti-French popular insurrection was the already mentioned Spanish War of Independence from the French of 1808–1814, which quickly became known abroad as the *Guerrilla*. After Ferdinand (VII) had forced his father, the Spanish Bourbon King Charles IV, to abdicate, Napoleon replaced both by his brother Joseph Bonaparte, causing resentment among the Spanish, who generally supported Ferdinand. The ensuing French military occupation with innumerable transgressions and atrocities against the Spanish population resulted in loyalist uprisings. They were originally decentralised and some occurred spontaneously or with the blessing of regional administrations (*juntas*) that had formed in the course of the uprising, but were soon coordinated centrally by the provisional government (*Junta Central*), and its successor regime, the *Cortes*. The *Junta Central*, and thus the uprising, was not recognised by the French, who treated the insurgents as rebels, in no way protected by the laws of war or conventions with regard to the treatment of prisoners.

The problem with the Spanish War of Independence as model for the new paradigm of 'small war as insurgency' is that it must not be reduced to this aspect alone. First, Charles Esdaile has shown that the Spanish uprisings of 1808 must be understood against the background of much older organisation of local militias to keep invaders from the north at bay during the previous centuries of Franco-Spanish War. These had been known under various names – as *somatenes* in Catalonia, as *alarmas* in Galicia, and as *migueletes* in Basque provinces. Collectively, these decentralised local security forces, which had seen much active fighting against the French in 1793–1795, were known as the *resguardo*. Importantly, then, the Spanish were used to mobilising the people in border regions to keep French armies at bay, and in 1808 there were many who themselves had experience in doing so; in resurrecting these formations, the Junta did nothing new or revolutionary.[26]

Secondly, contrary to the tenacious myth about the Spanish *Guerrilla* (which originated at the very time it was happening, as the contributions to this special issue show), this war was not fought exclusively or even mainly by peasants, townsfolk, and a few monks who had never previously raised a weapon. Instead, physical opposition to the French occupation came from a vast array of Spaniards who could be put into three overlapping and fluid categories, plus Portuguese and British regular forces fighting alongside them or separately. The myth of the Spanish *Guerrilla* has some substance in that there were indeed some town- and countryfolk who rose up to oppose the French, as Esdaile concedes.[27] But more of the fighting was carried out by irregular troops, recruited locally, trained, and commanded by experienced officers who in turn reported to and accepted the authority of the Junta. Called small war troops (*la tropa de guerrilla*[28]), parties (*partidas*) or free corps (*cuerpos francos*) or volunteers (*voluntarios*)[29] were clad in (usually brown or Hussar blue) uniforms, and fought and lived like the 'parties' under the *anciens régimes* – only, they were ideologically motivated, unlike Louis XIV's Dragoons or Maria Theresa's Hussars. They thus differed from seventeenth- and eighteenth-century 'parties' in that they were largely recruited from the populations of those areas where they were fighting, indeed, claimed to represent them, and claimed to be defending their own property, their own families, and village communities. Nevertheless, like their seventeenth- and eighteenth-century predecessors, they received regular pay.[30] Terminological clarity is further confounded by the fact that later in the nineteenth century the Spaniards would use the term *guerrilleros* in the eighteenth-century sense of 'parties', i.e. special forces or detachments, employed like regular soldiers directly by the government, and subject to considerable drill.[31]

Thirdly, among the Spaniards fighting the French we find regular soldiers in regular units of the Spanish Royal Army, which had resisted being disbanded. For example, at the Battle of Bailén on 21 July 1808 such forces defeated French General Dupont who had to surrender together with 18,000 French soldiers. The division between these (otherwise not very successful) regular forces and the partisans was fluid. On one occasion, an army (the seventh) was actually formed out of a collection of guerrilla bands, and on other occasions, cavalry units, like the 1000 men under Julián Sánchez García originally recruited to support an army, became more independent in their operations.[32] The tactics of the fighters of the second and third category were often the same. By 1812, most guerrilla bands had evolved into small divisions of regular troops. Some partisan leaders or *cabecillas* (heads in the sense of captain, French *chef*) like Sánchez García were or had been officers in the regular armies; others hailed from all walks of life from members of the landed establishment to bandit leaders.[33] Leaders, rank and file often moved between different categories, or rather, they changed their behaviour or status so that we would reassign them to different categories.

The Spanish War of Independence was the first of its kind in Europe since the religious and peasant wars of the sixteenth and early seventeenth centuries, as it involved such large numbers of irregular forces, and thus served as a model and inspiration for other uprisings against Napoleon. Yet the tactics followed by the *partidas* and the local self-defence forces were those of the *petite guerre* – constant harassment rather than pitched battles. Like the partisans of small wars of the eighteenth century, the Spanish did not win this war through small war tactics alone, but the decision was brought about by successive battles with British-led regular armies, culminating in their victory over the French at the Battle of Vitoria on 21 July 1813 under the Duke of Wellington. This would be an enduring feature of the people's war and all local insurrections against foreign military regimes in the future: they tend not to be decisive on their own, but only in combination with decisive conventional battles between regular forces fought alongside them. Whether this is always the case is a perennial debate, which can be traced back to the sixteenth and seventeenth centuries, as several of our contributions show.

Summary

To sum up, it is useful to divide 'small war' into two main categories. The first is that of special operations conducted by special detachments or specialised units operating alongside a normal field army or 'regular forces', a category mainly, but not exclusively, found before the French Revolutionary and Napoleonic Wars. The second is 'small war' in the sense of popular insurgency or 'people's war', the term coined by Heinrich von Brandt and/or Carl von Clausewitz.[34] The dividing lines between these categories – partisan warfare and insurgency – are blurred, particularly where tactics are concerned. The irregular forces fighting alongside the regular forces in the American War of Independence can be described both as partisans (special forces) and insurgents. In the Spanish *Guerrilla* of 1808–1814, organised local militias, partisans, spontaneous insurgents, plus the remnants of the Spanish regular army, plus regular British forces, plus regulars from countries allied to Britain like Portugal, fought side by side. As Sylvie Kleinman shows with her contribution, rebels could play the role of special forces (what the Spanish would call 'fifth column' a century later) operating for foreign states. Just as the British tried to use the Chouannerie in Brittany for their purposes against the French Revolutionary government, the French tried to engender anti-English Chouannerie-style uprisings in Ireland, South-Western England, and Wales.

It is not surprising, therefore, that there is a great confusion with regard to terminology, *guerrilla* being applied at different times to both sorts of 'small war', different writers using a term to denote different things. Words themselves have changed their meaning, so that over time *partisan*, originally in the eighteenth and early nineteenth centuries denoting the leader of a party of irregular soldiers, came to mean every member of the party, or the Spanish term '*guerrilla*' coming to denote the fighter in such a war (which in Spanish would normally be *guerrillero*, a term only commonly used even in Spanish from the late nineteenth century), sometimes dropping an 'l'. Other words crept in, often linked to particular technology and organisation, such as the 'riflemen' of the American Wars of the eighteenth century, the *tiradores* of nineteenth-century Spain, the '*tirailleurs*' of the French Revolutionary Wars, and the '*francs-tireurs*' of the Franco-Prussian War 1870–1871, 'snipers' in the twentieth century. Particularly after 1945, many insurgencies (especially anti-colonial uprisings) had a substantial Marxist input, and with this a myth of a *guerrilla* tradition of 'people's war' against oppressive governments and classes was fostered, giving both the terms *guerrilla* and partisan warfare a connotation of people's war which the latter term lacked completely until the anti-Napoleonic uprisings. Against this historical background, one is thus puzzled by ahistorical, theoretical treatments of 'partisan warfare' such as that of Carl Schmitt, which are incomprehensible outside a narrowly Cold War context.[35] The attention such distorting theoretical treatises have since been receiving is surprising.

Distortions are frequently a part of historical accounts of small wars, however, especially when the counter-insurgency forces emerged victorious and retrospectively committed to history their version of events, more often than not dismissing the insurgents as criminals. Those opposing bands of guerrilla fighters liked to call them bandits or *banditti*, an originally neutral word[36] that became widely known throughout Europe with the Italian *Risorgimento* (the various insurgencies and wars of national unity, 1815–1861), or *brigands* (as the French called Algerian insurgents in the 1830s). *Banditti* were members of *bande*, special detachments or units with a long-standing separate tradition of the sort which Benjamin Deruelle traces at least to the sixteenth century if not before, one of them headed, for example, by the dashing Italian nobleman Ludovico de' Medici, known as Giovanni dalle Bande Nere (1498–1526).

Ludovico de' Medici's fate should serve as a reminder that the distinctions we have drawn here for the sake of conceptualising the subject of this special issue are not always clear cut in reality. Ludovico can be regarded as a *condottiere*, a gun for hire together with his forces to the highest bidder, fighting the French on behalf of the Empire or the Empire on behalf of the French. His campaigns, however, which eventually led to his death from a wound received at the Battle of Governolo, could be summarised as a series of attempts to keep both the French kings and the Habsburgs out of northern Italy – and have been presented as an early manifestation of an Italian national consciousness.[37] Or one might consider the German knight Florian Geyer von Giebelstadt (1490–1525), chief of a band of *Landsknechte*, i.e. mercenaries, serving under the Ansbachs. In the Peasants' War of 1525, he first advised and then assumed the leadership of the insurgent peasants of the Tauber area, and was eventually killed for this political engagement. Even professional 'partisan' leaders could thus be inspired by values beyond mercenary considerations, just as ideologically driven insurgents could be motivated by the spoils of brigandage and crime, and many other personal motivations besides, as Charles Esdaile demonstrates in his contribution in this issue. In just the same way today, insurgents more often than not become embroiled with criminal networks, for want of other sources of arms or other vital supplies. In turn, professional special forces may be motivated by more than mere financial considerations or the lust for adventure. The admiration they receive from the general public if their exploits, where successful, become known must also be a factor.[38]

Acknowledgements

To conclude this brief introduction, it behoves me to thank the National Army Museum in London, which has made it possible for us to bring together several of the contributors to this special issue and to push them to produce their papers in English. Thanks are due especially Dr Alastair Massie, who organised the conference and was heavily involved in its intellectual development. Thanks are also due to Professor Hervé Drévillon of the University of Paris I Sorbonne and the École Militaire, with whom I organised an earlier conference on the topic in Paris in 2012. We are extremely grateful to the editors of *Small*

Wars and Insurgencies who encouraged us to produce this special issue and gave us invaluable help through peer reviews. Last but not least, I would like to thank Spyridon Plakoudas for his editorial assistance with this special issue.

Notes

1. For early examples, see Hahlweg, *Guerilla: Krieg ohne Fronten*; Ellis, *Short History of Guerrilla Warfare*.
2. Anon. (a Prussian officer), *Abhandlung über den kleinen Krieg*, 279.
3. Grandmaison, *La Petite Guerre*, 6–8.
4. We know this in part from the account of his grandson: Stuart, *Traité sur l'Art de la Guerre*.
5. Grandmaison, *La Petite Guerre*, 7f., 395.
6. Ibid., 14.
7. Ibid., 14f.
8. Ibid., 187.
9. Ibid., 149, 300–50 *passim*.
10. Ibid., 131.
11. Rink, *Vom Partheygänger zum Partisanen*, 124.
12. Ville, *De la Charge des Gouverneurs des Places*; De la Croix, *Traité de la petite guerre*.
13. Picaud-Monnerat, Sandrine. 'La "guerre de partis"', 202–34.
14. Zedler, *Grosses vollständiges Universal Lexicon*, Vol. 26, cols. 1049–50.
15. Jeney, *Le Partisan*, 1f.
16. Rink. *Vom Partheygänger zum Partisanen*; Picaud-Monnerat, *La Petite Guerre au XVIIIe Siècle*.
17. Valentini, *Abhandlung über denkleinen Krieg*, 1.
18. For the history of such units as part of the Empire-building of the early nineteenth century, see Frémeaux, *De quoi fut fait l'empire*.
19. Hargreaves, 'The Advent, Evolution and Value of British Specialist Formations in the Desert War, 1940–43'.
20. Duffy, *The '45*, 80–108.
21. Forrest, *Soldiers of the French Revolution*.
22. On Guibert, see Heuser, 'Guibert (1744–1790)'. For excerpts from Guibert's writings, see Heuser, *The Strategy Makers*.
23. Archduke Charles, 'Das Kriegswesen in Folge der französischen Revolutionskriege', 209.
24. Goethe took part in the campaign against France, and although he was not personally present at Valmy, he later claimed to have said at the time: 'From here and today a new era [Epoche] of world history begins, and you can say you have been there.' See Goethe, *Kampagne in Frankreich*, entry for 19–20 September 1792.
25. See Alan Forrest's contribution in this issue.
26. Esdaile, *Fighting Napoleon*, 29, 44.
27. Ibid., 43f.
28. Ofarrill, *Instruccion que deben seguir los oficiales y tropas*.
29. Carvajal, *Reglamento para las Partidas de Guerrilla*.
30. Esdaile, *Fighting Napoleon*, 38.
31. See Anon., *Instruccion de guerrilla* and Fernández de Córdova, *Tactica de guerrilla*.
32. Esdaile. Fi*ghting Napoleon*, 44–50, 53–7.
33. Ibid., 93.
34. Heuser, 'Small Wars in the Age of Clausewitz'.
35. Schmitt, *Theory of the Partisan*.

36. Note that powers fighting against insurgencies tend to dismiss them as illegitimate by categorising them as terrorists and criminals. While there tends to be an overlap especially with the latter, not least because insurgents generally find it difficult to obtain arms in states holding monopoly of force, this is obviously also a way of delegitimising the insurgents and dismissing their grievances.
37. Arfaioli, *The Black Bands of Giovanni.*
38. Note the standing of the SAS in Britain after the Iranian Embassy siege in London in 1980.

The sixteenth-century antecedents of special operations 'small war'

Benjamin Deruelle

Institut de Recherche Historique du Septentrion, University of Lille, France

The first conceptual, theoretical treatises about small war (*la petite guerre*) as special operations appeared only from the middle of the seventeenth century. The term is not used in the eighteenth-century sense of 'special operations' in older sources. The supposed absence of any treatment of the subject is surprising considering the obsession with the 'art of war' in the Renaissance, but other authors attribute it to a supposed antinomy between chivalric ideals and irregular warfare. But the absence of explicit manuals on the subject is not evidence of absence of advanced reflection on this kind of operations in the Middle Ages and in Early Modern times. We should thus look elsewhere, in other genres, for writings that contain and pass on military knowledge. Epics, romances, educational and military treatises, and memoirs in fact contain elements of a theory of special operations, even though these genres differ from our conception of rationality inherited from the Enlightenment.

The sixteenth-century roots of special operations 'small war'

The eighteenth-century strategist Capitaine Thomas-Antoine le Roy Grand-maison in his famous treaty about 'small war' (*petite guerre*) defined the objectives of 'parties' (special units) in such operations to include intelligence, small harassment operations, and attacks on small enemy units not least to extract booty, to find food and fodder, and to exact war subsidies from the local populations.[1] On the basis of other works of the seventeenth and eighteenth centuries, we might add to this any operations on the fringes of battles and sieges to assure the main army's security by closely observing enemy movements, to provide it with food and fodder while – if the opportunity arose – depriving the enemy of it, to harass the enemy, and to launch small attacks on small enemy contingents.[2] In the seventeenth and eighteenth centuries, small war consisted of skirmishes and limited operations of small detachments of regulars or irregulars that were members of a garrison or a field army, who could not decide wars by themselves, but made their contribution to the final outcome of a campaign.[3] This overlaps with what today one might call special operations, to distinguish it from small war (Spanish *guerrilla*) in the later sense of ideologically motivated popular insurgency or 'people's war' (in the language of the Prussians Heinrich von Brandt and Carl von Clausewitz, both writing in the 1820s[4]). This article deals with the question of whether 'small war' in the modern sense of 'special operations' existed before the term *'petite guerre'* began to be used in this seventeenth- and eighteenth-century sense.[5]

In the sixteenth century, troop contingents were generally smaller, even in battle. The difference between these special operations and 'regular warfare' (battles, sieges) lay neither in numbers nor the specialisation of soldiers, but in missions characterised by limited aims, surprise, tricks, stratagems, initiative taken at low levels of command, and constant adaptation.

Although these kinds of operations can be traced back to Antiquity, the first theoretical or conceptual works about them appear only from the Thirty Years War.[6] Yet military memoirs of the sixteenth century are replete with stories about them.[7] Moreover, the early modern period is driven by the obsession of the 'art of war'[8] born from the influence of Antiquity and a growing tendency to conceptualise the world in mathematical terms[9] that resulted in a rich harvest of theoretical works about warfare. In addition, France seems to have been the precursor in this domain with the manuals of Antoine de Ville, the Duke of Rohan, Maurice de Saxe, and Grandmaison in the seventeenth and eighteenth centuries.[10] The late arrival of works on the *'petite guerre'* is often attributed to the supposed antinomy between the *'petite guerre'* and earlier ideals of chivalry.[11] So was there no earlier general reflection on the conduct of these special operations? Perhaps scholars have looked in the wrong places.

This essay will turn to other genres or literature, namely epics such as *Galien Rhétoré* (1500), educational treatises such as *Le Jouvencel* (c.1460), general military treatises such as Raimond de Fourquevaux's *Instructions on the Feats of War* (1548), and military memoirs such as Blaise de Montluc's in the quest for such reflections and generalisations.

Special operations in the sixteenth century

The expression *'petite guerre'* was used in the sixteenth century, but we cannot find any dictionary entry for it before the second half of the seventeenth century,[12] nor can we find *'guerre de partisans'*, *'guerre de parti'*, *'aller en parti'*, or *'aller à la découverte'*. In the French dictionary of Robert Estienne and Jean Nicot dating from 1549 we find the term *'faire des courses'*[13] to mean to go out 'to the enemies, and plunder what one find on fields'.[14] And if we find the words *'parti(s)'* or *'partisan(s)'*, it is only in 1680 that we find it related to special forces (in the entry *'guerre'*) in the dictionary of Pierre Richelet, for whom a 'party' is 'a unit of soldiers that one sends to plunder, or to reconnoitre and ravage the enemy country'.[15] Gilles Ménage seems to be the first to use term *'petite guerre'* in this kind of literature in 1650.[16] However, it does not appear at the item 'guerre' but *'picorée'*:[17] 'Going to the *picorée*, is going to the *petite guerre*; to rob enemy of cattle, horses, sheep: what soldiers call a cow-run' (*'courre la vache'*).[18]

Definitions matching the eighteenth-century idea of the small war appear only in the second half or the last third of the seventeenth century, as in the preview published of the *Académie française* dictionary in 1687: 'to go to the war. That is to go in a party against the enemies'.[19] At this time, the expression *'petite guerre'* merits its own subentry under *'guerre'* besides others such as 'just and unjust war', 'civil war', and 'holy war'. Antoine de Furetière lists *'guerre'* and *'parti'*, and here we can find a definition that foreshadows Grandmaison's:[20] a detachment is a recognised *'parti'* if it has a 'written order from the commander, and if counts at least than twenty foot soldiers or fifteen horsemen, or else they are considered brigands'.[21] This reflects French government efforts to regulate *'petite guerre'* by positive legislation, such as in 1675.

Its absence from dictionaries does not mean that this form of warfare was unknown in the sixteenth century. In the late seventeenth century Gilles Ménage claims that Gilles Pasquier before him mentioned the expression *'petite guerre'* as a neologism of his times.[22] In the late nineteenth century, in efforts to trace obsolete words of the French language, Frédéric Godefroy and Edmond Huguet identify several synonyms or related terms, such as *'estradiotz'*, *'movant'*, *'paleter'*, and *'argoulet'*.[23] Before the seventeenth century, the term was frequently used in a different sense, however. Philippe de Commynes, in his *Chronicle of Louis XI's Reign*, denounces those who make a *'petite guerre'* to 'levy money', that is to say to tax people.[24] By implication this means a conflict begun for little reasons, vengeance or reprisals, and for small objectives.[25] In the Middle Ages, the term was used to denote private warfare, feuds between families or neighbours characterised by short, seasonal campaigns, skirmishes, and looting. These were campaigns limited in time and space, carried out by fighters who might be acting on some lawful authority, but not necessarily the central power (king or emperor). The term is used in this sense even at the end of the sixteenth century, as we can see in the translation of the Claude de Seyssel's *Peloponnese War* in 1527, or the *Political and Military Discourses* of François de la Noue in 1587.[26] It corresponds to the Spanish definition of the word *'guerrilla'* in the 1611 dictionary of Sebastián de Covarrubias.[27]

We also find the terms *'petite guerre'* used to denote behaviour that is not only undisciplined but immoral. In his correspondence, Agrippa d'Aubigné uses it for the behaviour of bad soldiers 'accustomed to steal hens [*courir la poule*] and do what the *argolets* (light horsemen) of this century named *la petite guerre'*.[28] Here, the phrase is a synonym for *picorée* or pilfering, a kind of plundering prohibited by the military authorities.[29] This ambiguity persists until the eighteenth century.[30] We have just seen that the first definition of *'petite guerre'* appears under the entry *picorée*, and Pierre Richelet in 1680 picks out the homonymy: 'This word is used when talking about war and is obsolete and is no longer used in good style, nor in ordinary conversations among men of the sword.' So in his time, one says undertaking the *petite guerre* and no longer, undertaking a *picorée* (*aller à la picorée*), and we also find the entry '*Picoreur*: Obsolete for one undertaking the *petite guerre*'.[31] Antoine de Furetière confirms and underlines that 'one called, in warfare, pilfering [*la Maraude*], this *petite guerre* that soldiers make by absenting themselves from the camp, going without orders or a leader to pillage from the peasant'.[32] In this sense, the phrase is used to define non-statutory actions that the crown tried to fight for a long time. However, the line between *petite guerre* and pilfering is often unclear in practice.[33] And the excesses of the Thirty Years War are not foreign to the bad image that the *petite guerre* keeps in the eighteenth century.[34]

So these *picorées* or unauthorised pillaging forays were distinguished from authorised special operations. In treatises and memoirs, we find other terms employed for the latter: most frequently '*entreprise*' (enterprise), '*course*' (run), '*chevauchée*' (ride), and sometimes '*destrousse*' (despoiling).[35] Soldiers dispatched on such operations were given names to correspond to their mission: '*coureurs*' (runners), '*avantcoureurs*' (scouts), '*descouvreurs*' (discoverers), or '*estradeurs*' (reconnoiterers). Nicot's dictionary from the beginning of the seventeenth century explains, '*courir*', i.e. to run, means also 'to spoil a country by hostile excursions [*courses*]', to send foot soldiers, light horses, or heavy cavalry 'against the enemies'. Such 'enterprises' were carried out by *coureurs* (runners) or 'lightly armed horsemen who are detached from the army, to reconnoitre the countryside, plunder and spoil the enemy's country', or 'ride about one league ahead of the army to reconnoitre and spoil the enemy's countryside'.[36]

We can draw two interim conclusions. The term 'enterprise' for such operations echoes the spirit of the Renaissance, its promotion of the individual and of the *virtù* stemming from an enterprising spirit, from calculated audacity and mental agility.[37] It introduces a social and cultural aspect into these operations. Secondly, in normative definitions at least, these operations and warfare are not differentiated. However, the absence of theoretical literature on the subject is not evidence of the absence of any kind of reflection. Sixteenth-century writers were more preoccupied with the distinction between what behaviour is acceptable in war according to the laws of war and Christian morality than their colleagues of the seventeenth and eighteenth centuries, who wanted to define all aspects of war.[38] Indeed, military occupation and the experiences of frontier war,[39] confrontation with imperial light troops,[40] the use of detachments, but also rejection of their often excessive violence, can be found in all three centuries.[41] At any rate, a variety of works written about warfare in the sixteenth century contain information about this kind of special operation.

Absence of theorisation?

Historians often imply that the special operations of the sixteenth century are known only through anecdotes from military memoirs and chronicles,[42] and not from theoretical treatises in the Vegetian tradition.[43] In fact, there are other genres of literature in which we can find discussions of them, such as literary fiction and allegories. In the Middle Ages as well as the Renaissance, these were used for didactic purposes, to pass on truth considered as universal, and this not only in the domain of morality or religion.[44] Therefore, epics and romances capture good practice and rules of warfare. Young gentlemen could learn warfare from reading the Classics (Vegetius, Polybius, or Thucydides), from contemporary theorists (della Valle, Fourquevaux, De La Noue), but also from chivalric literature.[45] Stratagems and ruses of warfare feature prominently in epics, where there are also many stories of other irregular operations.[46] As the epic hero incarnates the collective virtues of the military nobility, his adventures, successes, and failures summarise for the readers military and universal principles, and these include those applying to special operations.[47]

The epic *Galien réthoré*, written up in the fifteenth century from older stories and first printed in 1500, illustrates this well.[48] The story of Galien, a young gentleman in search of his father Olivier – one of the twelve peers of France and a companion of Roland – is full of accounts of ambushes, surprise attacks, and skirmishes done by Saracens or Christian knights. For instance, after the death of Roland and Olivier at Roncevaux, Galien takes charge of the army to finish the campaign in Spain. In order to find food, he sends 'four thousand men to forage who put everything to fire and blood. And they put to death all pagans they could find, and brought back with them much booty such as cattle, sheep and other useful victuals.'[49] These men act on orders with their only objective being to feed the entire army and to deprive the enemy of resources. In the next chapter, Galien uses surprise. A detachment of scouts discovers a pagan camp. In order to obtain more information on them, Girard de Vienne sends in a spy. Instead of choosing one of his squires, he selects somebody who knows their religious customs and their tongue so well that he is taken to be one of them. Having learnt that these men have laid down their arms and are tired by a night's march, Galien assails them to 'steal their meat and take what they have'.[50] The surprise is complete and the pagans are butchered.

Thus literary fiction neither explicitly articulates principles of war, nor does it present a doctrine, let alone theories. Nevertheless, it contributes to passing down military expertise inherited from medieval times. It uses the method of the medieval *exemplum* rather than theoretical abstractions familiar to us since the seventeenth century.[51] Such popular collections of secular *vitae,* just as before them the saints' lives, contribute to a universal knowledge and encourage the reader to copy the hero's action – *mimesis*. General principles are sketched from the listing of similar stories, making memorisation easier while leaving room for variation from case to case. The readers are invited to learn from diverse examples (*exempla*).

In the epics and romances, just as in reality, knights themselves or their squires carry out these operations, more or less heavily armed, depending of the nature of their mission. So the lack of distinction between such special operations on the one hand and sieges and battles on the other, or of operational specialisation between soldiers assigned to either task, shows that a perfect knight has to know all aspects of war, and that chivalric ideal and operations by detachments are compatible. In addition, these special operations provide a greater chance to express individual value and prowess.

The *Jouvencel* of Jean de Bueil, written between 1461 and 1468 and first printed in 1493, shares these characteristics chivalric fiction.[52] This short narrative treatise consists of three parts, headed 'monastic', 'economic', and 'political', on the Aristotelian model. The reader is invited to follow the progress of a 'poor gentleman' who, by learning to discipline first himself and then the others, manages to climb from the status of man-at-arms to captain and then to the higher responsibilities of the state.[53] Jean de Bueil thus distinguishes three different levels of warfare. In the first two, special operations dominate. Through the narration, the author puts across his own empirical knowledge of warfare.[54]

He describes operations undertaken to seize resources of enemies, to break up a siege, but also to overcome logistical problems, and to keep up the troops' morale and honour, and to bring in revenue.[55] The readers learn not only to prepare, organise, and lead a detachment, but also how to take a town by ruse or to protect one's forces against ambushes during marches. Jean de Bueil distinguishes two levels of operation, one for the small and one for the 'medium enterprise' depending on the objectives and the practices of specific tactics. Small forays are to be executed by night by detachments of 20 to 40 heavy horsemen preceded by scouts, far from roads, stopping frequently to listen and observe, and strike by surprise.[56] Assaults on camps and towns needed more men, between 300 and 1200, and involve greater preparations. At this level, de Bueil introduces, if not a specialisation, at least an adaptation of the men and equipment to the mission. For this a good captain has to dispatch scouts, lookouts on foot to watch the enemy and make contact with the local populations, and 'light companions' to reconnoitre the countryside and watch the lanes, the crossroads, and the hedges.[57] He distinguishes the scouts and the '*gardigeurs*', who might engage the enemy in skirmishes, and the country-runners, who are charged specifically with foraging.[58] Finally, de Bueil stresses the virtues that make the Jouvencel a good leader: courage and valour but also wisdom, subtleties, prudence, and knowledge of the countryside; acting with a sense of moderation and humility but also with perseverance and constancy; an ability to listen to advice from others while staying on one's guard and secretive.[59] He also avoids vices such as arrogance, envy, and avarice that make men more interested in plunder than in military objectives and the common good.

Therefore, the *Jouvencel* paints a good picture of special operations without, however, differentiating between it and regular warfare. By contrast, this distinction appears clearly in the first part of the sixteenth century in the *Instruction on the Actions of Warfare* of Raimond de Beccarie de Pavie, Baron of Fourquevaux, first published in 1548. This handbook is one of the most accomplished military reflections of early modern France.[60] Fourquevaux, a general and a diplomat, in his book not only describes the practice of his age but

23

puts forward proposals for reforms inspired by examples from Antiquity and the feats of the great captains of his time. He divides warfare into parts – recruitment, exercise, marches, camp, battle, pursuit, retreat. In this, special operations are presented as an auxiliary science. His book keeps some features already present in the *Galien* and in the *Jouvencel,* such as the confusion between special operations and regular warfare, the subjection of all operations to the goals of war, to manoeuvres, and the troops' general welfare.[61] Derived from examples, he articulates general rules for a good and wise captain.

Echoing classical authors, he recommends that a good captain should daily organise small operations to keep up the morale of the troops, their reflexes, and give them experience. He should send scouting parties ahead of his troops, to reconnoitre the countryside, to put out feelers to the locals, and watch out for changes in the enemy's behaviour.[62] The influence of Antiquity, the will to encompass all aspects of war and the spread of theories and practices favouring attrition, let Fourquevaux write about forced contributions and the plundering of the enemy's territory in order to achieve campaign objectives,[63] that is to say the use of violence against the civilian population, contrary to Christian morality.[64] In *Galien* and the *Jouvencel*, this violence is directed at pagans or is justified by extreme necessity, but for Fourquevaux it becomes a practical way to achieve war aims. For this reason, Fourquevaux praises the way Anne de Montmorency defended the Provence against the imperial invasion of 1536: Montmorency ordered his soldiers to 'destroy the mills and the ovens, and to spoil all the fruits that they cannot be saved [carried away], and pollute the fountains and wells ... : that act has been praised by both parties as the surest advice that could be applied at the time.'[65] Thus Montmorency deprived the imperial armies of food and water as they invaded France. Indeed the food shortages and dysentery forced Emperor Charles V to abandon the siege of Marseilles and to withdraw his troops.

What is also original about Fourquevaux, compared with his predecessors, is his designation, not only of foot soldiers but also of horsemen for these tasks.[66] The infantrymen are to fight, 'resisting everyone and smashing and breaking anyone they assail'; the light horses are to support fast advances, the scouts, and the search parties; the *stradiots* and mounted harquebusiers are to form scouting parties, night watches, and surprise operations. This distribution of tasks reflects the development since the fifteenth century of the light cavalry in Western armies on the oriental model.[67] Part of Fourquevaux's general project of military reform, this meant different career steps in the French army for gentlemen. To increase the heavy cavalry's efficiency, he wanted young gentlemen, after having served as pages, to attend 'little school', then to be harquebusiers for two or three years, then *stradiots*, and then to serve in the light cavalry (no sooner than at the age of 17) before being allowed to join the heavy cavalry on the model of knights of old. Only then would they have learnt to control 'the fire of their youth', having become 'cold and restrained for having learnt how to conduct themselves with wisdom among the other men-at-arms'.[68] So, like the author of *Galien* and Jean de Bueil, he asserted that nobles had to master all parts of war.[69]

24

These three examples show, first, the great variation of the writings designed to help nobles develop their military abilities. Even if they do not distinguish between general warfare and special operations, and even if the evidence is scattered, they treat all the principles, practices, and skills that are needed for this kind of operation to succeed. Only in the eighteenth century would treatises be written that would bring them all together in one work.[70] Moreover, the literature considered shows how perfectly irregular operations fitted in the military practice of French noblemen as a step in the introduction to the profession of arms. Nor were these practices at odds with the spirit of chivalry, as they converged on many points: booty, ransom, honour, and liberty were part of the code of chivalry.

Social uses of special operations

We have just seen that the *Galien*, the *Jouvencel*, and the *Instruction* present special operations as an integral part of the social and cultural representation of the military nobility. These see courage as a way to evaluate a person in a 'philosophy of action', inherited from the medieval epics and from Aristotelian moral conditions where prowess is considered the visible sign of military virtue.[71] Courage and prowess assume the social function of revealing and proving the nature of the combatant and his nobility (both in the sense of inherent quality and his membership of a caste) by demonstrating his skills and his legitimacy to command and govern the others.[72] They justify all quests for material and symbolic rewards as these can change the social standing of men and families. Therefore combat is understood as a qualifying ordeal, a way to distinguish oneself in nobility and as a consequence to rise in the social and military hierarchies.[73] By demonstrating his *virtù*, a warrior takes his destiny in his hands and challenges *fortuna*.[74] The stories supported by military memoirs attest that war and in particular special operations are considered as a means of distinguishing oneself.

The professionalisation of the army, the transformation of the use of heavy cavalry, and an army's daily needs make special operations the most frequent form of warfare.[75] This gives it an important place in the nobility's social representation. The Jouvencel and his companions are in garrison to toughen up and to learn by 'practical operations'. Daily *entreprises* protect them against idleness and keep their courage alive to be 'ever ready to use their body for honour and to acquire renown'.[76] His success gives him the opportunity to obtain promotion and his men's loyalty by sharing the spoils of war with them.[77] Beyond providing booty and spoils as a way of enrichment, the *entreprise* has an important social aspect. The pressure on young gentlemen engaged in it is extremely strong, because their performance in battle is essential to distinguish themselves. Their families, companions at arms, and chiefs watch out for any sign of courage or cowardice. Their first steps into the profession of war mark them, their name, and their families. Because it builds their *fama* (reputation), it determines their military career and their social standing. Blaise de Montluc, a

Gascon captain of modest extraction who rose up in the period of the Italian and French civil wars to become Marshal of France in 1574, told his readers 'to close their eyes to all perils and hazards of the first [military] encounters . . . ; because they are being watched to see if they have a strong stomach'.[78]

He himself applied these principles especially to members of his own family since their behaviour could also impact on his own reputation. In 1543, he led a troop of harquebusiers in an ambush of the imperial army in Piedmont, to the south of Turin. Upon encountering difficulties, he called on the Company of de Termes,[79] in which his 'nephew', Serillac,[80] was serving. Blaise de Montluc addressed him in this way: 'Serillac, you are my nephew, but if you do not attack the first, I disown you and say that you are not of my family'.[81] The young man obeyed, was among the first to charge, and had his horse killed under him by seven harquebus shots.

Imperatives of courage and honour weighed heavily not only upon young nobles who still had to prove themselves, but even on experienced knights.[82] Indeed, success in this kind of operation could make all the difference in the eternal competition between noblemen, bringing honour, recompense, and promotion within the social or military hierarchies. In case of failure, casualties, the loss of expensive equipment, and the need to pay ransom would compound the shame.[83] This led Blaise de Montluc like many of his contemporaries to urge caution with regard to military operations, submitting them to a thorough cost-benefit evaluation before engaging in them. In 1535, after Francisco Sforza's death, Francis I of France claimed the duchy of Milan and invaded Savoy. In reaction, Emperor Charles V launched the aforementioned invasion of Provence, reached Aix-en-Provence, and besieged Marseille. The supply of flour and bread for his imperial troops was provided by mills around Arles. One in particular directly catered to the needs of 6000 soldiers.[84] This mill lay five leagues inside the area controlled by the Spaniards and was defended by 60 men and the garrison of the town. Its destruction was thus a perilous enterprise that no captain wanted to undertake. And in 1536, no captains of Francis I wanted to take it on. Even Montluc decided to volunteer only after noting the king's displeasure at the lukewarm reactions of the other captains. This episode is interesting, first, because it is presented as a model of *entreprise* done to deprive enemies of supplies;[85] secondly, because it is often given as a counter model of epic and heroic action.

Montluc did not differ from his peers and superiors because of his immoderate taste for risk, his particular bravery, or his military talent, but because of his lack of renown and honour. He was fully aware of how risky a business he was about to embark upon.[86] However, after suffering an injury 1528, he had stayed away from the army for three years and felt that this long absence had made a big dent in his prestige. As a consequence, he had to 'start all over again . . . as an unknown person seeking my fortune at great peril to my life'.[87] We can suppose that he might have refused this mission before 1528 when his reputation was well established, but now it was the occasion for him to 'make myself known to the king, and get back to my old reputation and standing with the great that I had previously'.[88] This anecdote shows what part special operations could play in the process of legitimising a social rise based on courage, prowess, and merit, and by the chivalric culture.[89] The bitterness of competition among noblemen and the pressure on gentlemen to prove their worth, in order to retain respect and justify daily their place in the military hierarchy, make these operations on the fringe of the great battles and sieges a particular means of expressing military prowess.[90] We see similar patterns in Shakespeare's Histories, from *Richard II* to *Richard III*, where even the peers of the realm were under pressure to perform well in war, and owed their promotion to highest office to their bloodline but also to their exploits in war. All this explains why anyone would '*aller faire le carabins*',[91] another expression for this enterprise, in spite of all dangers.

Conclusion

To conclude, we have seen, first, that there were sixteenth-century antecedents of what in the eighteenth century was called *petite guerre*, and that reflections on it were not absent from the military literature of the sixteenth century. The texts we identified for the late fifteenth and sixteenth centuries that dwell on '*courses*' and '*entreprises*' show a consciousness of this kind of combat and its perfect fit with noble culture. And even if epics, romances, and military treaties do not make any distinction between regular warfare and special operations, they do establish rules and a framework of reference for contemporaries in ways different from ours. Rather than producing a linear evolution towards the theorisation of these special operations, the sixteenth and the beginning of the seventeenth centuries saw a transformation of the way in which people thought about knowledge. If only Fourquevaux seems to fit in with our system of rationality, conditioned by the Enlightenment, the distinction between these special operations and regular warfare is expressed more clearly in the *Jouvencel*. At last, the social structures and the nobility's habits and ethos have to be considered to go beyond a purely military understanding of special operations. Because in a system of representation dominated by knighthood and chivalry, special operations furnished opportunities to gain merit, and demonstrate prowess and honour, notions that would still be recognisable in the following centuries when embarked on the '*petite guerre*'.[92]

Notes

1. Grandmaison, *La petite guerre*, 111. See also Beatrice Heuser's Introduction in this issue.
2. Lynn, *Giant of the Grand Siecle*, 538–42; Couteau-Bégarie, *Traité de stratégie*, 238; Peschot, 'La notion de petite guerre en France (18ᵉ siècle)', 137; 'La guérilla à l'époque modern', 8; Picaud-Monnerat, *La Petite Guerre au XVIIIe Siècle*, 25–9.

3. Lynn, 'How War Fed War'; Ravaille, 'La petite guerre dans les commentaires de Monluc', 6.
4. For this distinction, see Heuser, 'Small Wars in the Age of Clausewitz'. On the debate over the use of 'small war' and 'small wars' in English, see Picaud-Monnerat, *La Petite Guerre au XVIIIe Siècle*, 36–40; '"Partisan Warfare", "War in Detachment"'.
5. Peschot, 'La petite guerre au 16e siècle', 262; Picaud-Monnerat, 'La "guerre de partis"', 136.
6. See, for example, Grandmaison. *La petite guerre*, 1–4; and the analysis of Picaud-Monnerat, 'La "guerre de partis"', 102; and Couteau-Bégarie, *Traité de stratégie*, 238.
7. Peschot. 'La petite guerre au 16e siècle', 262.
8. Burckhardt, *La Civilisation en Italie au temps de la Renaissance*, 140; Verrier, *Les armes de Minerve*.
9. Koyré, 'Galilée et Platon', 170.
10. See, for example, Henri duc de Rohan. *Le parfait capitaine*; de Ville, *De la charge des gouverneurs des places*; de Saxe, *Traité des légions*; de Folard, *'Des Partis de guerre'*, in *Nouvelles découvertes*.
11. Deruelle, 'De papier, de fer et de sang', ch. 1.
12. *Grand Corpus des dictionnaires [9e-20e s.]*. The conclusions presented here come from the systematic study of the forms identified by Ludwig Ravaille, Sandrine Picaud-Monnerat, and Bernard Peschot as inheritance from the sixteenth century, and from my own observations in military memoirs.
13. Literally 'do runs'.
14. Estienne, *Dictionnaire François latin*, 145; Nicot, *Thresor de la langue francoyse*, 159.
15. Richelet, *Dictionnaire françois*, 126. We also find in the first edition of the *Dictionnaire de l'Académie française*, 2: 188.
16. The phrase is in the *Origines de la langue françoise* of Gilles Ménage but not before the 1694 edition (1st ed. 1650).
17. Synonym of marauding coming from the verbal form to peck.
18. Ménage, *Les Origines de la langue françoise*, 574.
19. *Dictionaire de l'Académie françoise*. Paris, Petit, 1687 [Avant-Première 3]; and in *Dictionnaire de l'Académie française*, 1: 546.
20. Grandmaison, *La petite guerre*, 3.
21. Furetière, *Dictionnaire Universel*, 2: 214.
22. Ménage, *Les Origines de la langue françoise*, 574.
23. Godefroy, *Le Dictionnaire de l'ancienne langue française*, 3: 636, 5: 433, 707; Huguet, *Dictionnaire de la langue française du seizième siècle*, 1: 299–300; Picaut-Monnerat, 'La "guerre de partis"', 113; Ravaille, 'La petite guerre dans les commentaires de Monluc', 8.
24. Commynes, *Chronique et histoyre*, f. 85.
25. Contamine, 'C'est un très périlleux héritage que guerre', 6, 7.
26. *Lhistoire de Thucydude Athenien*, f. v–v°; La Noue, *Discours politiques et militaires*, 529.
27. Covarrubias, *Tesoro de la Lengua Castellana o Española*, 666.
28. Aubigné, 'Lettre à M. l'evesque de Maillezais', 405–6, 417.
29. La Noue, *Discours politiques et militaires*, 684–6.
30. Picaud-Monnerat, 'La "guerre de partis"', 131–2; *La petite guerre*, 132–7.
31. Richelet, *Dictionnaire François*, 159.
32. Furetière, *Dictionnaire universel*, 2: 551; see also *'Picorée'* in 3: 118. The same distinction is made in Corneille, *Le Dictionnaire des arts et des sciences*, 4: 25; *Dictionnaire de l'Académie française*, 2: 22.
33. Lynn, *Giant of the Grand Siecle*, 544.

34. Picaud-Monnerat, *La petite guerre au 18ᵉ siècle*, 15; 'La petite guerre au 18ᵉ siècle en Europe', 180; Balvay, *L'épée et la plume*, 157.
35. Bueil, *Jouuencel*, f. v, v v°, vii; Anon. [Fourquevaux] *Instruction sur le fait de la guerre*, f. 48, 53 v°.
36. Nicot, *Le Thresor de la langue francoyse*, 57.
37. Delumeau, *La civilisation de la renaissance*, 336–7.
38. Picaud-Monnerat, *La petite guerre au 18ᵉ siècle*, 19, 29–40; Gat, *History of Military Thought*, 27–31; Lynn, *Battle*, 125ff.; Guinier, *L'honneur du soldat*.
39. Drévillon, *Les rois absolus*, 20–7, 215, 470 *et seq.*; Chanet and Windler, *Les ressources des faibles*.
40. Pepper, 'Aspects of Operational Art'; Antoche, 'Les guerres irrégulières dans les principautés de Moldavie et de Valachie (14ᵉ-15ᵉ siècle)'; Czigány, 'Tradition et modernité dans les affaires militaires du royaume de Hongrie'.
41. Balvay, *L'épée et la plume*, 150; Lynn, 'How War Fed War'; Peschot, 'Les lettre de feu'.
42. Couteau-Bégarie, *Traité de stratégie*, 238; Peschot, 'La guérilla à l'époque moderne', 3–12; 'La notion de petite guerre en France' 135–48. On the theorists of the eighteenth century, see Picaud-Monnerat, 'La "guerre de partis"', 105.
43. See Tallett, *War and Society in Early Modern Europe*, 39; Heuser, *The Strategy Makers*, 'Introduction'.
44. Strubel, '*Grant senefiance a*'; Pérez-Jean and Eichel-Lojkine, *L'Allégorie, de l'Antiquité à la Renaissance*; Lubac, *Exégèse médiévale*; Mora-Lebrun, *L'Énéide médiévale*, 97–108; Wolff, *Fulgence le mythographe*, 115–16; Pionchon, 'La Généalogie des dieux païens entre le Décaméron'.
45. See, for example, the prologue of *Galien rethoré*, f. 3–3v°. Chartier et al., *L'éducation en France du 16ᵉ au 18ᵉ siècle*, 177; Deruelle, 'Enjeux politiques et sociaux de la culture chevaleresque au 16ᵉ siècle'.
46. Deruelle. *De papier, de fer et de sang*, 210; see also Holeindre, *Le renard et le lion*, ch. 4.
47. Paquette, 'Introduction', 23–4; Gaucher, *La biographie chevaleresque*, 519–20; Poulain-Gautret, *La tradition littéraire d'Ogier le Danois après le 13ᵉ siècle*, 42.
48. Paris, 'Galien'; Horrent, *La chanson de Roland*, 377–412; *Galien le restoré en prose*, 11–16; Keller, 'Autour de Galien le Restoré'.
49. *Galien rethoré.* f. lvii.
50. Ibid., f. lviii.
51. Berlioz and Polo de Beaulieu, 'Les prologues des recueils d'*exempla*'; *Les exempla médiévaux Introduction*; *Les exempla médiévaux: nouvelles perspectives*.
52. On Jean de Bueil – gentleman of arms who reached the grade of admiral de France (1450) and the dignity of knight of the king's order of Saint Michel (1469) – and his book, see the introductions in Bueil, *Le Jouvencel*, ed. Tringant et al., i–cclxx16, cclxx19–cccxx12; Blanchard, 'Écrire la guerre au 15ᵉ siècle'.
53. Bueil, *Le Jouuencel*, f. vi°.
54. Blanchard, 'Écrire la guerre au 15ᵉ siècle'.
55. Bueil, *Le Jouuencel*, f. v et seq., xxx v°, xl v°, xl v°.
56. Ibid., f. xix v°, xx v°, xxxiii, xxxv.
57. Ibid., f. xxvii, xxvii v°.
58. The word '*gardiguer*' is absent from the edition of 1496, but appears in one of the medieval manuscripts. Bueil, *Le Jouvencel*, 149. La Curne Sainte Palaye emphasises this distinction but asserts that it was no longer used in his times, see his *Dictionnaire historique de l'ancien langage François*, 4: 330.
59. Bueil, *Le Jouuencel*, f. vi, vii, vii v°, x, 17 v°, xix v°, xx.

60. For excerpts in English translation, see Heuser, *The Strategy Makers*. On this treaty and its paternity, see Lefranc, *L'armée Française et la Renaissance*, 46.
61. Fourquevaux. *Instruction*, f. 48, f. 56, 65 v°.
62. See, for example, *Flave Vegece Rene*.
63. Tallett, *War and Society*, 59–60; Pepper, 'Aspects of operational Art', 181–2.
64. Vauchez, 'La notion de guerre juste au Moyen Âge'; Hubrecht, 'La guerre juste dans la doctrine chrétienne'; Mattéi, *Histoire du droit de la guerre*, 1: 407–9; Deruelle, 'Des limites imperceptibles'.
65. Fourquevaux, *Instruction*, f. 52 v°.
66. Ibid., f. 23.
67. Fratani, 'Les chevaux des Gonzagues'; Szabó and Tóth, *Mohács (1526)*, 73–81; Czigány, 'Tradition et modernité dans les affaires militaires du royaume de Hongrie', 272–8.
68. Fourquevaux, *Instruction*, f. 23.
69. Verrier, *Les armes de Minerve*, 57–61; Peschot. 'La petite guerre au 16ᵉ siècle', 263.
70. Lynn, *Giant of the Grand Siecle*, 543–6.
71. Aristotle, *Nicomachean Ethics* II.1; Deruelle, *De papier, de fer et de sang*, 165.
72. Jouanna, *L'idée de race en France*, 69; *Ordre social*, 120–36; Storrs, 'The Military Revolution and the European Nobility', 33.
73. Febvre, *Honneur et Patrie*, 83; Pitt-Rivers, 'La maladie de l'honneur', 20; Drévillon, 'L'âme est à Dieu et l'honneur à nous'.
74. Pocock, *Le moment machiavélien*, 35–55; Skinner, *Les Fondements de la pensée politique modern*, 138–155; Verrier, *Les armes de Minerve*, 57.
75. Pepper, 'Aspects of operational Art', 182.
76. Bueil, *Le Jouuencel*, f. viii v°.
77. Ibid., f. xxx.
78. Montluc, *Commentaires*, 1: 337.
79. Paul de la Barthe governor of Savigliano.
80. Serillac is in fact the son of Jean de Serillac a Blaise de Montluc cousin.
81. Montluc, *Commentaires*, 1: 482.
82. Jouanna, *Le Devoir de révolte*, 63; Drévillon, *L'impôt du sang*, 321–442; Ihl, *Le Mérite et la République*, 55; Deruelle, '"Pour Dieu, le roi et l'honneur"'.
83. This recalls the 'shame culture' identified by Benedict, *The Chrysanthemum and the Sword*; Dodds, *The Greeks and the Irrational*.
84. Auriol is located 30 km north-east of Marseilles in the direction of Aix-en-Provence.
85. Tallett, *War and Society*, 59; Harari, *Special Operations in the Age of Chivalry*, 163–81; Pepper, 'Aspects of Operational Art', 197.
86. Montluc, *Commentaires*, 1: 402.
87. Ibid., 1: 385.
88. Ibid., 1: 388.
89. Deruelle, *De papier, de fer et de sang*, 188; Arnaud Balvay has underlined the concordance between petite guerre and chivalric war of the Middle Ages (*L'épée et la plume*, 150).
90. Drévillon, 'Courtilz de Sandras'.
91. Peschot. 'La petite guerre au 16ᵉ siècle', 262; 'La guérilla à l'époque moderne', 5–6.
92. Gaurat, *Journal de Simon Delorme*, 94, 106–7, 109–10; Lynn, *Giant of the Grand Siecle*, 539; Drévillon, *L'impôt du sang*, 385–91.

The essence of war: French armies and small war in the Low Countries (1672–1697)

Bertrand Fonck and George Satterfield

Service historique de la Défense, France; US Naval War College, Newport, USA

In the late seventeenth century during the Dutch War (1672–1678) and the Nine Years War (1688–1697), French armies relied on small war for the accomplishment of essential tasks and as part of an overall strategy of exhausting their opponents in the Low Countries. The purposes of small war included the imposition of contributions on enemy populations, the destruction of the enemy base of operations, blockades of fortresses, and the general support of campaign armies. The expression 'small war' in the French language appeared with growing frequency in the 1690s. Small war can be viewed as both a cause and consequence of the characteristics of these wars. The limited policy goals of Louis XIV the king of France required a strategy that minimised risk and accomplished the goal of reducing if not eliminating the Spanish presence in the Low Countries that bordered the north of France. As French armies increased in size during this period, the demand for specialists at small increased in order to provide security and ensure supply. Small war in the late seventeenth century was thus not ideologically motivated insurgency, but in the minds of French commanders an essential component of strategy and the nature of war.

As the introduction to this special issue has shown, in late seventeenth-century Europe, the term 'petite guerre' or 'small war' principally described a variety of actions of war that took place in conflicts between sovereign states. It turned up frequently in reports by commanders, official correspondence, and memoires. If the manoeuvres of the armies of the period were slow and ponderous, combat was nevertheless a daily event and the cumulative effect of the numerous and violent clashes of small war sometimes determined the outcome of entire campaigns. Courtilz de Sandras testified to the intensity of small war in his *History of the Dutch War* (1689) when at the beginning of the year 1676: 'the war blazed so much from one side to the other that winter, which is customarily a time of repose for soldiers, differed not all from the campaign, except that one sometimes returned to recover in their winter quarters. They were nearly always mounted, they conducted sieges, and fought in skirmishes, and these were particularly bloody as they were drawn at closer quarters than most battles.'[1] Yet, surprisingly, the subject of small war in the late seventeenth century has remained largely underappreciated. The classical representations of the great battles of Condé and Turenne and on the one hand, and Vauban-style sieges on the other, endure deeply etched on the imagination. Compared to the numerous volumes dedicated to the major operations of the period, scholars have only recently begun to evaluate the place of small war.[2]

Our argument here, based on research in the French army archives at Vincennes, is that within the theatre of war of the Low Countries small war constituted the predominant effort by French armies in both the Dutch War (1672–1678/9) and the Nine Years War (1688–1697). Military historians who have focused on battles and sieges in these wars have arguably missed the bigger picture formed by small war. The tasks of small war in these conflicts were vital to success in war and they included: (1) contributions or war taxes; (2) raids of various types; (3) blockades; and (4) reconnaissance, convoy security, and other operations directly related to campaign armies. A growing demand for specialists at small war, called partisans in the sources, and the perspective of the Court and army commanders, all point to the general importance of small war in the evolving strategies of Louis XIV that spanned two major coalition wars.[3]

Contributions

Parties or small detachments of troops were constantly solicited for the imposition of contributions. Parties of troops distributed leaflets of demand, collected payments, and conducted punitive expeditions against recalcitrant villages, which often resulted in smouldering ruins and the taking of hostages.[4] Parties took up the role of terrorising populations in order to force them to pay and to ensure an orderly collection of contributions. They also worked to enforce the deliverance of passports, which made possible the taxation of travel and thus commerce in occupied territories. Parties also served as safeguards in those villages that had paid contributions and an additional wartime tax for from the depredations of soldiers from both sides. It was certainly due to contributions and the related war taxes of passports and safeguards that civilians most often interacted with armies, whether passively in their payment or more aggressively in the organisation of local militia and armed resistance.[5]

The relative weight of small war in the late seventeenth century was crucially related to the importance attached to the imposition of contributions, which became more organised and a priority under the eyes of Louis XIV's minister of war Louvois. During the Dutch War in the Low Countries, contributions and related war taxes reached the amount of 13 million *livres* or French pounds a year from 1674 to 1678. This amount came to about 16% of the annual extraordinary expenditures used to prosecute the war.[6] The goods taken in kind as contribution payments, often forage or grain but sometimes livestock, or paid in coin, served primarily to supply the French garrison forces, but when necessary could be used to support campaign armies, allowing French armies to take to the field before their adversaries extending the duration of campaigns and allowing for more ambitious strategies.

At the beginning of his reign, Louis XIV and his war ministers, the father Le Tellier and later his son Louvois, rationalised the imposition of contributions. A series of royal ordinances centralised the process and placed it under civil control. Intendants of the crown, directly responsible to the minster of war, supervised where contributions would be imposed, the amounts, and whether detachments of French troops might be necessary to enforce their payment. Fortress governors and army commanders were left with the military responsibility of dispatching forces as directed: 'to carry out those executions which will be required of them by the aforesaid intendants'.[7] When the Dutch War spilled over into the Spanish Low Countries in 1673, the process was essentially in place for the remainder of the Sun King's wars.

Two distinct French forces operated in a coordinated fashion throughout the wars of Louis XIV, and this was especially true of French efforts to impose contributions. One force consisted of the campaign armies capable of *grande guerre* or the major operations of war, but dispersed into winter quarters for five or six months of the year. The other force was composed of the fortress garrisons in the theatre of the Low Countries. During the Dutch War, both forces were relatively equal in size, each averaging around 70,000. The garrison force supported campaign armies in a variety of ways, providing security for convoys, gathering intelligence, and conducting raids against enemy foragers. But through their own efforts campaign armies also met these tasks of small war and on occasion even supported the garrison force in the imposition of contributions. Contemporaries understood these actions of war, including the imposition of contributions, as aspects of small war; *petite guerre* entered the military lexicon about 1690, earlier it was known as the 'war of parties', or the 'war of partisans', among other colourful, descriptive phrases.[8]

The formalisation of the system of contributions thinly disguised its deeper nature, which was no less than the law of the strong. The refusal to deliver the contributions demanded, following the customary reminders, and armed resistance by villagers toward the collectors normally led to military reprisals called 'executions'. These operations of controlled destruction, which in most cases involved the burning of houses or of entire villages, took place within a broader context of destruction carried forth by small detachments. Louvois gave encouragement to his army commanders in this regard, writing to the Marquis de Chamlay in September 1676 to 'lose no opportunity to torment all the country there, until the inhabitants have entirely satisfied what they owe'.[9] In September 1690, during the Nine Years War, he exhorted the Duke of Luxembourg: 'If one puts the torch to several houses or villages in that country, the others will soon be shaken as well.'[10]

The executions for contributions did not always fall on defenceless peasants, but sometimes encountered more formidable defences manned by armed militia and regular troops. In these cases, the 'typical' detachment designated to conduct an execution might include 100 if not 1000 or more French troops. Such was the case repeatedly when the French attempted to impose contributions on a small but relatively distant corner of the Spanish Netherlands called the 'Waes country' between Antwerp and Ghent. In October 1675, Marshal Humières led a raid that included 800 cavalry, 3000 infantry, 4 light pieces of artillery, and 2 pontoon bridges. Spanish regulars and peasant militia were routed from their entrenchments and more than 1200 houses were burnt to the ground. In another large-scale execution on the Waes, on 26 April 1677, the equivalent of a cavalry colonel in Louis XIV's army, *mestre-de-camp* Ezékiel Mélac, informed the French intendant at Maastricht of a large-scale execution that destroyed the village of Kekern and 'two leagues around... thoroughly pillaged and nearly all burned'.[11]

A cycle of reprisals inevitably followed in the small war to impose contributions as each side pursued the payment of contributions in the theatre of war. Louvois, in particular, became more and more implacable in the cruel enterprise of reprisals. In January and February 1684 during the brief War of Reunions between France and Spain, nearly 160 villages were burnt as reprisal in the Spanish Low Countries, all the way to the gates of Brussels. It becomes evident that the famous devastation of the Palatinate (1688) largely found its roots in the widespread practice of imposing contributions and the brutal retaliatory raids that had ensued in the Dutch War and the War of Reunions.[12] Louis XIV himself assumed this policy after the death of Louvois in 1691 in requiring that in the Nine-Years' War (1688–1697), 20 Spanish villages be burnt for every French village that had been destroyed.

Raids

The purposeful harassment and destruction of enemy defences further placed small war at the forefront of the scene. Raids for purely destructive purposes, as opposed to executions for contributions, targeted the physical base of operations for the Spanish, Dutch, and imperial armies in the Low Countries. The targets were frequently defences manned by garrisons of regular troops ensconced in outlying castles or small towns protected by antiquated walls. In practice, French detachments would assault virtually any place that the adversary might find useful as shelter, for winter quarters, for the storage of supplies, or of use in providing some control over the surrounding countryside and that might prevent the payment of contributions. Raids altered the physical geography of entire swaths of countryside in the Low Countries, as when the walls of the town of Visé on the Meuse River were demolished following its capture in January 1675, and soon afterwards, the governor of French-held Maastricht resolved to demolish 'all the castles which are in the vicinity of the Meuse... in order to make the passage from Maastricht to Liège, entirely free'.[13] The following spring in 1675, a French campaign army led by Louis XIV in person completed the destruction carried out by the garrison at Maastricht with the conquest of Huy and Dinant on the Meuse River. In January 1676 the garrison of French-held Maastricht undertook two raids simultaneously in order to remove garrisons of regular troops that hindered the imposition of contributions in Liège. The total force involved in this raid included 600 cavalry, 200 dragoons, 2000 infantry, two light cannon, and a petard to blow open gates. However, after a reconnaissance party reported that the defences of one of the towns had been considerably strengthened, that assault was called off. The other attack on the enemy garrison in the town of Theux went forward and broke its defences.[14]

Some raids aimed at demolishing defences and destroying bases of operation were quite daring in concept and involved the escalade of walls at night, the swimming of moats, and the crossing of rivers by means of pontoon bridges brought along by the raiding force. In the same month that the French garrison of Maastricht captured Theux, another raid struck a castle outside of distant Bonn, which served as the capital of the Electorate of Cologne and that had prevented the surrounding lands from paying contributions. A lieutenant of dragoons with 25 men swam one moat of the castle, swam a second moat, and finally entered the castle lowering the drawbridge for the rest of his command to enter. The castle was plundered and hostages taken to ensure that the region paid the contributions that it owed.[15] The destruction of the twin castles called Ecaussines resembled sieges only on a small scale. Marshal Humières was entrusted with the capture of the castles and organised a force that included 8000 troops, infantry, dragoons, cavalry, and two siege guns – the entire expedition was undertaken by garrison troops in the dead of winter. Both castles were 'invested', according to the normal procedures for a formal siege. The first castle attacked, after light resistance, surrendered within a day. The attack on the second castle required sappers to advance an assault trench forward and its defenders withstood over 'a hundred volleys' from the siege guns.[16]

Numerous raids in the Low Countries focused on the supply infrastructure used by the enemy. Many of these were conducted against river barges that supplied enemy garrisons, as when a French party burnt in January 1676 two barges carrying forage and other supplies to enemy strongholds on the Scheldt River.[17] Villages near enemy fortresses were in particular hard hit, since they often sheltered supplies for the immediate use of a nearby garrison. Around the village of Oudenaarde in July 1676, French parties hauled away over 300 wagons and carts; forage that could not be carted away was wasted 'so that the enemies will not be able to help themselves to it'.[18] Louis XIV encouraged French raids for a more immediate political purpose, as when he personally advised a French commander to 'ruin the country of Cologne, as much as possible'.[19] The intention was to punish the Elector of Cologne who had once served as French ally but who had switched sides in 1677.

Propagandists of both camps used the destruction of raids to cast the other side in a negative light. The French destruction of Bodegraven and Zwammerdam in Holland in 1672 and the destruction of the Palatinate in 1689 were but two more infamous examples that propagandists in the coalition camp used against Louis XIV. Yet neither side attempted to conceal the brutality of raids. In a French almanac for 1691, side by side with illustrations of the victories of the French campaign army in 1690 was placed an illustration of a violent raid conducted by the Comte de Tessé against the duchy of Jülich.[20] The intention of the raid illustrated in this work of propaganda was clearly to demonstrate the futility of minor powers that dared to resist the all-powerful forces of the Sun King.

The day-to-day destruction carried out by raids in enemy lands and the execution of villages in the gathering of contributions had a pronounced effect in weakening the morale of the adversaries of Louis XIV and in undermining the support of their populations. Following the destruction of the two castles called Ecaussines in 1676 for example, the major landowners of Spanish Brabant were 'totally outraged' and some personally requested 'neutrality for their castles' from the French Marshal Humières. Others instigated a riot against the Spanish governor-general that killed two Spanish soldiers in Brussels.[21] In 1678 in order to coerce the Duke of Palatinate-Neuburg to cooperate with Louis XIV, Luxembourg planned to destroy one of his private châteaux, as he explained to Louvois: 'I think that it is better to demolish it than to burn a dozen of his villages, because he does not worry about the harm to his lordship and this [personal] approach will grab his attention.'[22] Perhaps the most spectacular result in terms of the moral effect of a raid took place in 1675 when French parties based at Maastricht raided deep into Holland and purposefully burnt homes belonging to members of the Dutch government. The Dutch government urged the commander of its armies, Prince William, to lay siege to Maastricht, an operation that ended disastrously for the Dutch in 1676.[23]

Small war required a continuous coordination of all the actors, and its cumulative effects exceeded by far any single action of the campaign armies. The complexity of small war alongside major operations also promoted the establishment of what scholars have called 'cabinet strategy'. Thus the war of parties, only in appearance secondary, formed an essential part of a single, coherent strategic vision that was essentially one of exhausting the opponents of Louis XIV. Here we can understand better the limited objectives that Louis XIV provided his campaign armies, and his reluctance to risk the fate of battle, since what mattered in his eyes was the reduction of the Spanish presence in the Low Countries and the sparing of his own lands. The parallel of the strategy of exhaustion on land with the evolution of French maritime strategy with its increasing preference for commerce raiding as opposed to major fleet actions easily comes to mind.[24] Small war was thus far more than a tool for the support of armies in the field and of siege operations – it was an essential element of strategy. The Court's frequent refusals to reinforce campaign armies with troops from the garrisons involved in the small war, and that opposed the wishes of the commanders of campaign armies, found its justification in the overall strategy of exhaustion.

Blockades

Blockades of major fortresses composed yet another aspect of small war and served the strategy of exhaustion by limiting the duration of sieges and reducing the risk of battle that might result from an extended siege. The process of blockades followed a discernible pattern: detachments of French troops cleared defenders from villages and country castles around fortresses months in advance of a formal siege; parties fortified the advanced posts with earthworks; they turned the bell towers of churches into observation posts; patrols were mounted to prevent supplies or reinforcements from reaching the blockaded fortress. During the Dutch War, French strategy emphasised the blockades of Cambrai, Valenciennes, and St Omer from 1676 until they were formally besieged in 1677. Blockades should not be construed as passive operations, especially if one weighs the plight of civilians into the balance and the risks undertaken by French parties. Parties of French troops actively engaged in seeking out and destroying any forage or supplies in the immediate countryside surrounding a blockaded fortress. The French commander Baron Quincy explained in a letter to Louvois in June 1676 that the objective of the blockading force at his disposal was no less than to force the Spanish and their allies to abandon their provinces in the Low Countries: 'Without a doubt, *Monseigneur*, the consumption of the grains of this country will cause this winter a terrible scarcity among the enemies, who otherwise cannot to find the money to purchase it; thus left deprived of their bread, you take away the use of their weapons and in consequence their country.'[25] That same month Baron Quincy's blockading force fought in two cavalry engagements against Spanish defenders from Cambrai and Valenciennes. The second battle resulted in the Spanish losing more than 400 soldiers, including 27 officers. During the winter, which proved exceptionally cold, parties of French troops were instructed to make obstacles out of ice to block fords that might be used by the enemy to bring supplies into the blockaded fortresses.[26] A conspiracy was set in motion to burn the magazines within blockaded Cambrai along with several buildings that housed the municipal government.[27] Another French commander, the Comte de Bulonde, would write in February 1677, outside of blockaded Valenciennes that the French post at 'Crespin is a cruel obstacle to them, nothing can enter the plain of Valenciennes without the [sentries] in our bell towers seeing it.'[28]

Both Louvois and Louis XIV fully endorsed the patient strategy of blockades and exhaustion, although the king's preference was for more elaborate operations, full-scale sieges, as he explained to Louvois: 'After examining all of the alternatives all of the alternatives that you have proposed to me, I believe that there are no others ... the big sieges please me more than the others, but in the state of matters, it is necessary to postpone to another time.'[29] The Sun King did not have long to wait. Once the garrisons of the major Spanish fortresses were isolated and worn down for want of supplies, instead of sieges taking 41 days (the average length of a siege following a formula by Vauban),

Valenciennes withstood only 16 days of siege, Cambrai 26 days, and St Omer 14 days. All three sieges were concluded in the summer of 1677, and blockades of Mons and Ghent would continue into the next year. In January 1678, Louvois made it clear that in order to reduce the garrison of Mons parties of French troops should 'make a desert of the villages two or three leagues around the town', and to 'make as many disorders' to force the inhabitants to 'leave their homes'.[30]

The preference for the blockade as part of an overall strategy of exhaustion, and the small war necessary to make it happen, continued even after the peace of Nijmegen in 1678 as Louis XIV attempted to 'straighten out' the frontiers of his newly conquered territories at the expense of Spain. Louis opted for a strategy of raids and the occupation of minor places since he desired to avoid formal hostilities and another coalition formed against him so soon after the conclusion of the Dutch War. A cold war took place between the Dutch War and the Nine Years War. In the region between the Sambre and Meuse rivers, partly in preparation for a renewed war with Spain over the political destiny of the Low Countries, detachments of French troops occupied 'in the span of two years ... three fortified towns, five abbeys, twenty-one castles, and nearly two-hundred villages', without any reference to a conference that was established by the Treaty of Nijmegen to regulate such matters.[31] No major operations were conducted during these 'reunions'. French small war operations in this interval of peace, however, did lead to open hostilities with Spain in the War of Reunions, 1683–1684, during which the strategy of blockades was continued: Spanish-held Luxembourg was blockaded from January to April 1684 before a formal siege began. Louis XIV's preferred strategy of exhausting his opponents blurred the boundaries between war and peace during the interval between the Dutch War and the Nine Years War.

Small war and campaign armies

Alongside the struggle for contributions and the strategy of raids and blockades, small war was paramount for campaign armies that increasingly depended on small war in the accomplishment of their principal operational goals: the conduct of formal sieges and the neutralisation of opposing armies by battle if necessary. The gathering of intelligence for campaign armies was one of the principal missions of small war and of specialist partisans who served as the eyes of campaign armies. Parties effectively constituted one of the favoured means possessed by a general for the gathering of information on an enemy position, the strength, and the movements of an opposing army. Combining speed with discretion, responsiveness, and mastery of terrain, parties were able to infiltrate the protective screen of opposing armies or, by the conduct of daily patrolling, catch unawares and take prisoner enemy troops detached to provide security for a convoy or a forage party. In some cases, French parties departed for several days in total autonomy; in other cases, they acted closer to home as the advanced guards of the field army while on the march. In order to multiply their chances of success and to be advised of the least movement of the enemy during the Nine Years War, Marshal Luxembourg commonly sent five or six parties simultaneously, which regularly sent information as much by oral reports as by the dispatch of written reports. The closer two opposing armies were, the more frequent the reports and the more rapid the transmission of intelligence. Luxembourg noticed for example in September 1690: 'I am near enough the enemy that they cannot make a march that I will not be warned three hours after.'[32] The interception of enemy couriers without a signed passport that allowed their free passage was yet another means for reconnaissance parties to draw intelligence. Parties would take prisoners from enemy forage parties, groups of marauders, deserters, advanced guards, and from enemy reconnaissance parties encountered; the interrogation of prisoners provided valuable information to general officers who showed a preference for intelligence gained in this direct way. Yet the struggle for intelligence that surrounded every campaign army led to frequent and bloody encounters with the enemy. In his dispatches, Luxembourg mentioned with regularity that parties returned with prisoners (and valuable information) while on other occasions he related that enemy detachments found face to face had not been given any quarter.

A good partisan from a campaign army established reliable contacts with the local population that allowed him to obtain first-hand information from another source, and there were also those partisans who disguised themselves as peasants for missions more related to espionage than to military operations. Spies helped partisans in the small war to disrupt the supply sources of opposing armies. In 1676 Louvois insisted that men be sent 'from all directions in order to be secretly informed of places where they [the enemy] will make depots of forage or of grain'.[33] An example of the respective roles of spy and partisan can be found in the hours leading up to the battle of Steinkerque in 1692. William III succeeded in a perfect deception operation by unmasking a spy of Luxembourg on the eve of the battle and in making him communicate to Luxembourg that the allied army was about to conduct a large-scale forage operation. Yet it came about that, as result of the service of a reconnaissance party and its urgently repeated messages, Luxembourg was finally convinced that the alleged forage was in reality a deployment for battle.

Other critical missions fell to parties during the wars of Louis XIV: from an offensive perspective the attack of enemy detachments, foragers, and convoys and, from a defensive perspective, the security of camps, marches, foragers, and convoys. It was essential to avoid surprise attacks, and even if the armies camped generally in the disposition of the order of battle, every effort was made in the choice of positions and in covering the approaches to the camp by detachments to ensure security. But it was always possible for adventurous and daring parties to elude this cordon and approach sufficiently near to attempt the capture of men or horses. It was not unusual for the most capable to even succeed in entering an enemy camp, provoking a general panic. Louis XIV delighted to learn in 1691 of the success of a party in this delicate exercise, regretting however that it had been unable to return with any captured enemy flags.[34]

Partisans set off as scouts during army marches, in which they would be tasked with covering the different marching columns that armies formed. During the Dutch War, campaign armies typically formed into four or five columns. By the 1690s, the larger armies commanded by Luxembourg in the Low Countries generally separated into at least five or six columns, and sometimes up to ten, which followed more or less parallel routes; it was thus all the more necessary for detachments to secure the multiple routes of march. An attack on the columns of an army on the march was generally beyond the reach of partisans, but such was not the case for the foragers and convoys that went back and forth from encamped armies. An attack mounted against foragers from the enemy camp was the daily bread of the partisan, who might just as equally be called upon to serve as part of a cordon to protect a friendly forage operation. If one cannot consider all the detachments from armies sent to forage as war parties, especially when sometimes an entire wing of an army was tasked with a major forage operation, parties were nevertheless indispensable for the protection of foragers that were tempting and valuable targets for the partisans of both sides. The activity of parties also ensured security against threats to a campaign army from further afield.

The protection of convoys was another valuable service performed by parties: large convoys, bringing the subsistence of a magazine to army in the field, and that might include several thousand wagons; and smaller convoys of artillery trains or of army baggage trains. In these missions, the French sometimes sustained stinging setbacks as recounted by the journal of the writer Dangeau, who described rather lengthily in his memoires a combat which cost the life of the Count de Verillac, a brigadier, who in 1693 made an attempt to defend a convoy of 700 wagons against an enemy party from the garrison of Charleroi.[35] Campaign armies could face considerable difficulty as a result of such attacks on convoys. During the Dutch War, in July 1674 a convoy bringing sacks of flour to Condé's army was ambushed and destroyed as the covering escort was caught riding in the wagons.[36] Conditions worsened in the French camp and as a result officers even encouraged the men under their command to maraud in the surrounding countryside for food, a practice reminiscent of the Thirty Years War and something not at all approved by the Court. When a Spanish party ambushed another convoy sent to relieve Condé's army on 10 July, 3500 sacks of flour were lost. By these persistent partisan attacks, the Spanish forced Condé's army to abandon its position and opened the way for the advance of their own army into French territory.[37]

Whether offensive or defensive, the action of partisans led them to encounter their counterparts of the opposing side frequently. Thus combats often of a reduced scale but sometimes relatively important (sometimes involving several thousand men from each side) took place nearly on a daily basis, which contradicts the alleged propensity of campaign armies in this period to avoid any form of confrontation. The randomness of the engagements caused unequal combats that were not always in favour of the side with the most troops. The actions of parties were in fact a theatre of the most direct and violent combat, often hand-to-hand struggles, but also of ruses and stratagems. Luxembourg recounted for the benefit of the Court in September 1692 the victory of a partisan, who, with 140 cavaliers, had dispersed two enemy squadrons, double his number, by beating a drum behind a hedge in order to simulate the arrival of reinforcements.[38]

The specialists of small war

The importance of the diverse missions of small war led to a rapid increase in the troops necessarily devoted to accomplishing its tasks. Small war involved varied actors and many talents but increasingly specialists. With respect to the effectives involved, parties were composed in general of several dozen men, but some counted several hundred or more as previously noted. There existed a lower limit by mutually agreed upon convention of 19 infantry or 15 cavalry below which parties were considered as 'blue parties' or marauders. The upper limit was not as precise since the difference between *petite guerre* and *grande guerre* resided in the nature of the mission assigned to the troops. Parties could thus be commanded by lieutenants or by general officers, according to their importance. In August 1692, Marshal Luxembourg sent Rosen, a lieutenant general, with 500 cavalry and 100 dragoons, toward the enemy; the party encountered the enemy and killed 50 in a particularly bloody action.[39] Even more substantial bodies of observation troops were frequently sent on reconnaissance and to take prisoners; to this end, Luxembourg sent 1400 men toward to the enemy on one occasion during the Dutch War.[40] Yet the nature of these actions, involving even hundreds of troops, fits fully within the definition of small war as understood by contemporaries.

The composition of parties naturally reserved a place of preference for mounted troops, but the cavalry was far from possessing a monopoly in the action. Parties of infantry were frequent in this era, providing a wonderful role for the all-purpose grenadiers, and even for mixed parties of cavalry and infantry. The era did not yet know light cavalry in the sense that it would be understood later, at least until the formation in 1692 of the first regiment of hussars, descendants of the Croats of the Thirty Years War. Marshal Luxembourg employed volunteers from Hungarian deserters, accustomed to the tactics and nuances of *petite guerre*, even before the creation of this unit; he profitably used his Hungarian volunteers to imitate the hussars which served in the imperial German ranks and that terrified and outclassed numerous French parties from the beginning of the Nine Years War.

But it is necessary to pause here concerning the essential role of dragoons, which constituted an arm apart from the cavalry. Mounted infantry served in the role of light cavalry: dragoons were perfectly equipped for *petite guerre*: supple boots, a *fusil* or musket, a brace of pistols, bayonet, and a sapper tool. Dragoons counted then among the privileged actors of small war during the wars of Louis XIV. One cannot avoid making a connection between the rapid increase of this arm in the years 1670–1690 and the increasing need for more troops devoted to small war. Dragoons reached their apogee in the last decade of the seventeenth century with a total of 43 regiments, of which 29 were organised between 1688 and 1690.[41] Marshal Luxembourg disposed in 1693 and 1694 of no fewer than 24 squadrons of dragoons, where they were employed in advance of marches in order to clear the way for the columns of cavalry[42] and to cover the baggage train.[43] He particularly used them as an escort when he wished to reconnoitre a position personally.

Dragoons and later hussars perfectly answered then the demands born with the spectacular increase in manpower between 1672 and 1697 and were tied to the increasing complexity of the manoeuvre of armies in multiple columns. Whether composed of infantry, cavalry, or dragoons, the practice was to organise parties not by companies or entire units but by combined detachments from several units, somewhat like today when special, joint task forces are organised for exterior operations. The volunteer was the rule when smaller parties and detachments were formed, and even if all troops could be used, some were more specialised or experienced in the rigours of small war and its special demands. Experienced and motivated partisans were found in both the garrison force and the campaign armies.

Essential to small war were various companies belonging to the governors and lieutenant governors of fortress garrisons, companies of dragoons and fusiliers that composed part of the garrison force but specialised more in the actions of small war than in the immediate defence of fortress walls. Armed and equipped in a similar fashion to the regiments of dragoons that normally served with the campaign armies, the governor's companies were reinforced for small war with men recruited from the surrounding region who were familiar with the roads, paths, and places suited for ambushes. They provided an offensive capability to the fortress governors and were essential for the collection of contributions, the security of convoys sent from fortress magazines to campaign armies, and to provide for a general defence against the actions of enemy parties.

Partisans found interest in their risky sorties, but also in the remuneration that permitted them to build a reputation, opening even the possibility of a career. There was an economy of *petite guerre*, which Hervé Drévillon has illustrated.[44] Partisans kept a share of the plunder that was put up for auction, and this was the main motivation for their daily risking of life. Nevertheless, small war was not always profitable, and partisans who lost their horses quickly found themselves broke; moreover, the risk of small war was a factor which accounts for the repugnance of many captains to contribute volunteers to parties, as well as the practice of composing parties from different units which allowed risk to be shared along with the rewards. Within the command of his campaign armies, Marshal Luxembourg did everything possible to protect the financial interests of his partisans, for whom he regularly requested bonuses. He proposed to the king in

1691, for example, that money should be provided to his best partisans for horses.[45]

French commanders like Luxembourg regarded their partisans as especially valuable assets and even took a personal interest in their better partisans, men who had set themselves apart by their courage in combat. Saint-Simon testified to the high regard that commanders held for their specialists at small war by writing in 1693 about the famous partisan Le Fèvre: '[Le Fèvre was] captain in our regiment, who, a former keeper of pigs, had arrived there by way of merit and of promotions, and who could neither read nor write, although an older man. He was one of the best partisans in the king's army, who never departed without seeing the enemy or reporting sure intelligence. We loved him, held him in high regard and considered him one of us, and so did the generals.'[46] Luxembourg confirmed this judgement, soliciting the preceding year some extra compensation for this deserving partisan. Saint-Simon wrote in the same way about a man named Tracy, who warned Luxembourg of the approach of the enemy army at Steinkerque: 'he became one of the best partisans of the army. His determination, his valour, the perfect execution of everything in which generals regularly entrusted him to accomplish, acquired their respect, then their favour. He earned the complete trust of M. de Luxembourg.'[47] The best partisans were in effect highly sought after; Luxembourg thus requested from Louvois to approve the arrival in the army of a Swiss named Matheis who had gallantly served in the garrison of Ypres.[48] He equally proposed to Louis XIV to agree to provide a company to a partisan who had switched sides and served Spain, so that he might be persuaded to return to the service of France.[49] In July 1691 the French army celebrated the arrival in its ranks of one named Dubray, known as the best Spanish partisan at that time.[50]

Yet most of the commands of parties went to officers that one might call generalists and not specialists. Officers of the *maison du roi*, among other elite units, often volunteered to participate in small war in a leadership role. If the destruction of villages and (presumably) the slaughter of the inhabitants appeared a little unsavoury to some French officers, the activity of the partisan was not yet discredited and was on the contrary valued, at least according to the correspondence of Luxembourg, who himself had been shaped by the 'war of posts' during the Fronde and the Franco-Spanish War in the 1650s. As in the previous century (as Benjamin Deruelle shows in his contribution to this special issue), *petite guerre* appeared to many young officers as a school to learn the arts of war. Marshal Villars who later fought Marlborough to a standstill at Malplaquet in 1709, having learned his lessons in the Dutch War and Nine Years War, could later write, 'nothing is more fitting to educate a true warrior than the experience of a craft that teaches him to attack with boldness, to withdraw with order and good sense, and lastly that accustoms him to seeing the enemy at close range.'[51] Louis XIV himself valued small war as an essential part of an officer's education. The king wrote to Luxembourg in 1692: 'You can never arouse the officers enough to practice themselves in that kind of war which contributes so much to their learning, and which is moreover as advantageous to them as it is damaging to the enemy.'[52]

Paradoxically, military writers like Feuquières, Puységur, and Chamlay who had participated in the campaigns of the Dutch War and the Nine Years War theorised very little on *petite guerre* or small war. The writing on small war in this era was a subject found only in passing in general treatises on war. Commanders nonetheless revealed a preoccupation with it in the official correspondence with the Court. Marshal Luxembourg attached great importance to it, and his correspondence gives abundant evidence of it, illustrating even the interest of the Court and of Louis XIV, who often immersed himself in its details. While true that the actions of partisans were indispensable to the security, mobility, and supply of armies, campaign armies had become in the 1690s so large that they became difficult to maneuvre. Luxembourg was fully aware of the functions that parties fulfilled and he regularly requested reinforcement by garrison troops and particularly by their partisans, without outwardly worrying too much about the weakening of fortress garrisons. As a general rule the larger the armies under his command, the more Luxembourg recognised the need to devote more troops to the relevant missions of small war in order to clear the way for the march of his armies and to protect his convoys. Luxembourg, when requesting reinforcements, often explained that his best troops were constantly overextended, occupied in forages, the guard of camps, the protection of convoys from enemy parties, leaving only weak, untested troops for greater offensive actions. He attempted nevertheless to preoccupy adversary forces with offensive actions in order to dissuade them from mounting their own offensives on the lands that belonged to the king of France. To his dismay, he was often forced to protect French territory by leaving a part of his own forces behind to guard the lines of field fortifications established in western Flanders to prevent the imposition of allied contributions on French soil. Thus, Luxembourg recognised that in a strategic sense small war and *grande guerre* were interrelated and inseparable. Perhaps it was this recognition, widely shared, that prevented a specialised theory of small war from being developed.

Success in small war was often viewed in moral terms as a kind of test of the overall capabilities of armies. Commanders spoke in terms of achieving a moral ascendancy over the enemy in small war the effects of which would exceed the result of any single engagement taken separately. Louis XIV wrote to his general in 1691: 'It is a good omen that the parties of my army always defeat those of the enemy, and there is reason to hope, that what they accomplish in a minor way, they will equally succeed in doing in a major way.'[53] Moral ascendancy, morale, is in an important subject and closely linked to the value system of officers who sought to enhance their personal reputation.[54] Small war also served another indirect purpose: it reinforced the values of honour and the individual bravery of officers, besides satisfying by small sips the king's thirst for glory.

At the level of the Court, small war formed part of a coherent strategy of exhaustion placed at the service of a policy of limited conquest in the Spanish

Netherlands. During both the Dutch War and the Nine Years War, the goal of Versailles was to push the enemy to a negotiated settlement on favourable terms. Operational objectives were moderated accordingly and were consistent with the defence-in-depth of both sides, lines of fortifications that both sides attempted to break, and the immense challenges in the supply of armies. In this scheme, small war played a role of first importance, of equal significance as battles, sieges, and the manoeuvre of armies.

The significance attached to small war in the later seventeenth century is undoubtedly part of the story of the evolution of war in general. If the partisans of Louis XIV are placed in the tradition of medieval raiding warfare that continued until the time of the Thirty Years War, the contrast is clear enough: small war more regularly fitted within an overall strategy during the wars of Louis XIV than in previous times when small war was more often the consequence of attempts of armies to merely survive from one campaign season to the next than part of a rational strategy. Yet the French practice of small war and change in the scale of its employment during the period 1672–1697 can be seen both as a cause and a consequence of the limited nature of war, since small war remained a crucial necessity and not an option that armies might pursue in particular circumstances.

Conclusion

We should emphasise the numerous contributions that the study of small war makes to our understanding of warfare in this period. First, it allows us to shed light on a daily practice of war and a fundamental dimension of strategy too often obscured by the *grande guerre* or major operations. It contributes as well to the debunking of the received notion that combat seldom took place in a period before the soldier became an automaton, as in the era of Frederick the Great, and the value of individual bravery was not yet suppressed by the culture of service and the economy of violence. It equally makes a connection between the role of field armies and that of garrisons, and the actions taken in the year-round the cycle of operations, from the months of winter quarters to the months of the campaign season. It also involves operations that directly placed in contact armies and civilians, a topic often set aside or ignored in the narratives of grand operations.

Small war was a particular response to the operational problems posed by war at a time when the resources of states were comparatively limited. It was not merely a matter of secondary or supporting operations as it was sometimes defined in the eighteenth century, but by its essential nature a war of harassment of the stronger forces by the weak – a clear manifestation of what is now known as asymmetric warfare. By the means of small war, the enemy could be defeated without the risk and costs associated with battle – although not without fighting. During the Dutch War and Nine Years War, small war was clearly integrated in a broader strategy on the same level as the actions of campaign armies where it remained until the end of the eighteenth century. By the 1820s and 1830s, it appears that small war had begun to be thought of in two diverging ways: as ideologically driven insurgency or as supporting or secondary operations to major war (*'grande guerre'*) – no longer strategically distinctive or coequal.[55] Carl von Decker's treatise on small war in French significantly bore the title 'small war or treatise of secondary operations in war'.[56] Direct payment by the state for forage, requisition systems, the diminished importance of contributions, and a moral critique of small war contributed to it being downgraded to secondary or supporting operations. The Jominian and Clausewitzian tenet built on the experience of the Napoleonic wars that wars were decided by victorious battle relegated to the sidelines any alternative views about the conduct or aims of wars.[57]

For the period we are dealing with here, we should set aside and finally break with the idea of classical warfare associated with the sequence of sieges and battles. The skirmishes of cavalry painted by Van der Meulen in his youth have as much place in the baroque fury of war in the late seventeenth century as the large compositions of the methodical sieges that he painted in his later life. Before small war gained its theoretical autonomy in the mid eighteenth century, the reign of Louis XIV served as a period of transition during which small war remained as it had been for a long time simply war – even the essence of war.

Notes

1. Courtilz, *Histoire de la guerre de Hollande*, 2: 1.
2. Lynn, *Giant of the Grand Siècle*; Peschot, 'La guérilla à l'époque moderne'; *La guerre buissonnière*; Satterfield, *Princes, Posts and Partisans*.
3. On the growing demand for partisans and the viewpoint of France's leading soldier of the 1690s, Marshal Luxembourg, see especially Fonck, 'Le maréchal de Luxembourg'.
4. Lynn, 'How War fed War'; *Giant of the Grand Siècle*, 184–217. The author speaks of a 'tax of violence' to designate the violent requisitions and plunder, progressively limited by the administration of the reign of Louis XIV, and replaced in large part by contributions. Peschot, 'Les "lettres de feu"'. The work of Jean-Pierre Rorive provides a precious study of the subject and the consequences of the actions of French armies at Huy and in its region in *Les misères de la guerre sous le Roi-Soleil*.
5. Chanet and Windler, *Les ressources des faibles*.
6. Satterfield, *Princes, Posts and Partisans*, 42–3.
7. Service Historique de la Défense [henceforth SHD], library, Recueil Cangé, Ordonnance du Roi, concernans les contributions... 23 September 1673, vol. 23, no. 21.
8. For example, the engraver Nicolas Guérard on the subject of parties in the series *Exercises de Mars*, published in 1695, utilised the expression *petite guerre* to describe their general purpose.
9. SHD, GR A^1 484, no. 294, Louvois to Chamilly, 7 September 1676.
10. SHD, GR A^1 943, no. 178, Marly.
11. SHD, GR A1, 537, no. 132, Mélac to Calvo, 26 April 1677.
12. Cénat, 'Le Ravage du Palatinat'.
13. *Gazette de France*, Liège, 7 February 1675.
14. SHD, GR, A^1 498, no. 17, Estrades to Louvois, 5 January 1676.

15. SHD, GR, A^1 486, no. 105, Estrades to Louvois, 12 January 1676.
16. SHD, GR, A^1 486, no. 301, Humières to Louvois, 26 January 1676.
17. SHD, GR A1, 486, no. 11, Chamilly to Louvois, 1 January 1676.
18. SHD, GR A^1 487, no. 89, Chamilly to Louvois, 27 February 1676.
19. SHD, GR A^1 533, no page, Louvois to Calvo, 2 August 1677.
20. Larmessin, *Les grandes victories*.
21. *Mercure Hollandois*, March 1676.
22. SHD, GR A^1 593, no. 60, 13 October 1678.
23. SHD, GR A^1 448, no. 20, Estrades to Louvois, 6 January 1675.
24. Cénat, *Le roi stratège*.
25. SHD, GR A^1 500, no. 262, Quincy to Louvois, 23 July 1676.
26. SHD, GR A^1 485, no. 313, Louvois to Bulonde, 29 December 1676.
27. *Mercure François*, December 1676.
28. SHD, GR A^1 536, no. 7, Bulonde to Louvois, 5 February 1677.
29. SHD, GR A^1 484, no. 203, Louis XIV to Louvois, 1 August 1676.
30. SHD, GR A^1 595, no. 19, Louvois to Montal, 3 January 1678.
31. Jeanmougin, *Louis XIV à la conquête des pays-bas espagnols*, 31–2.
32. SHD, GR A^1 943, no. 186, Luxembourg to Louvois, 5 September 1690.
33. SHD, GR A^1 485, no. 214, Louvois to Du Chaunoy, 22 December 1676.
34. SHD, GR A^1 1051, no. 24, Louis XIV to Luxembourg, Versailles, 4 September 1691.
35. *Journal du marquis de Dangeau*, 5–6 July 1693.
36. SHD, GR A^1 405, no. 173, Le Peletier to Louvois, 5 July 1674.
37. SHD, GR A^1 405, no. 186, Robert to Louvois, 10 July 1674.
38. SHD, GR A^1 1140, no page, Luxembourg to Louis XIV, 19 September 1692.
39. SHD, GR A^1 1139, no page, Luxembourg to Louvois, 5 August 1692.
40. *Gazette de France*, Charleroi, 26 August 1675.
41. Sarmant, 'Une seconde cavalerie', 233–5.
42. SHD, GR A^1 1047, no. 21, ordre de marche de l'armée pour le 19 mai 1691, partant de Curne sur la Lys pour aller à Hautrive sur l'Escaut.
43. SHD, GR A^1 1048, no. 52, ordre de marche pour le 9 août 1691, de Cerfontaine à Lugny prés Beaumont.
44. Drévillon, *L'impôt du sang*.
45. SHD, GR A^1 1048, no. 41, Luxembourg to Louis XIV, 6 August 1691, and A^1 1048, no. 43, 8 August 1691.
46. Saint-Simon, *Mémoires*, 1: 239.
47. SHD, GR A^1 1139, no page, Luxembourg to Louis XIV, 21 August 1692.
48. SHD, GR A^1 942, no. 5, Saint-Amand, 9 May 1690.
49. SHD, GR A^1 943, no. 223, Luxembourg to Louvois, 5 October 1690.
50. SHD, GR A^1 1049, no page, Bagnols to Louvois, 5 July 1691.
51. *Mémoires du Marechal de Villars*, 1: 13–14.
52. SHD, GR A^1 1138, no page, Louis XIV to Luxembourg, 22 September 1692.
53. SHD, GR A^1 1041, no. 39, Louis XIV to Luxembourg, 31 July 1691.
54. Drévillon, *L'impôt du sang*.
55. Heuser, 'Small Wars in the Age of Clausewitz', and see the Introduction to this issue.
56. Decker, *Der kleine Krieg*.
57. On the preponderance of the 'Napoleonic Paradigm', see Heuser, 'Victory, Peace, and Justice'.

Initiating insurgencies abroad: French plans to 'chouannise' Britain and Ireland, 1793–1798

Sylvie Kleinman

Department of History, Trinity College, Dublin, Republic of Ireland

Secret French plans to launch guerrilla-style raids on the British Isles devised in the spring of 1796 were referred to as 'chouanneries'. The name and concept behind these small-war operations were modelled on the irregular tactics used by the Chouan rebels in the Vendée, which the French state army had brutally quashed, but some wished to transfer into their institutional practice. Part of France's ongoing military strategy in the war against Britain, which included fomenting insurrection in Ireland, these irregular operations were to be manned partially by pardoned deserters and released convicts and prisoners of war. Of these, only Tate's brief invasion of Wales in 1797 was realised, but the surviving plans provide insightful historical lessons into an Anglophobic mindset shared by a small network of practitioners and policy deciders on the effectiveness of such shock and awe tactics. Largely motivated by the desire to take revenge for Britain's support of counter-revolutionaries in the Vendée, these plans could more aptly be referred to as counter-'chouanneries'.

Tate's Landing: The last foreign invasion of Britain

On 22 February 1797, the last foreign invasion of Britain occurred when a motley band of *c.*1200 French troops descended on a poorly defended spot of the Welsh coast in Fishguard Bay about 150 miles overland to Bristol.[1] This Black Legion, commanded by an exiled veteran of the American Revolution, Colonel William Tate, included some seasoned troops but many irregulars, deserters, released royalist prisoners, and even convicts.[2] Back in France Theobald Wolfe Tone (1763–1798), who was to become an iconic hero of Irish nationalism, had even derided them as 'banditti... and sad blackguards'.[3] (Tone was a secret negotiator who had arrived in Paris on 1 February 1796 and received a French commission in July.[4]) Despite outnumbering local forces, Tate surrendered after two days as many of his well-armed desperadoes had descended into drunkenness and plunder. The fiasco triggered a frenzied rush on cash worsening Britain's financial crisis, but also called loyal Britons to arms to prepare for further invasions while coastal defences were reinforced. Cherished in Welsh folk memory due to the bravery and patriotism displayed by civilians (including females in red cloaks), this odd occurrence of state-sanctioned irregular warfare is little known in the annals of history. There is a small plaque on the Welsh coastal footpath commemorating the landing,[5] but the whole episode is not widely enough known to challenge the popular claim that Britain has never been invaded since 1066. When referenced at all in more recent times, the episode has been succinctly written off by Frenchmen as an embarrassing reminder of a bygone age of both romantic and reckless campaigns, fuelled by revolutionary fanaticism, facilitated by divisive government, symptomatic of a weak navy, and often reliant on reprobate adventurers rather than professional soldiers. One French military author, Général Paul-Constant-Amédée Gastey, clarified he had *not* included the Tate invasion in his article on General Humbert's brief campaign in Ireland in 1798, as it was but 'a dishonourable affair led by a band of guttersnipes... which ended lamentably'.[6]

If ill-prepared, the British had been expecting such an attempt. It was undoubtedly convenient to interrogate their Anglophone prisoner, and hardly a surprise to learn that three of his junior officers were Irishmen. The previous December, a major French expeditionary force carrying close to 15,000 troops had nearly succeeded in landing at Bantry Bay on the south-west coast of Ireland, and was to have provided the military framework for an Irish insurrection.[7] Its commander was one of France's most intrepid generals, Lazare Hoche, who had repulsed the British-backed landing at Quiberon (Brittany) in June 1795 and comprehensively 'pacified' the counter-revolutionary *Chouans* in western France.[8] Hoche's fleet had been dispersed by ferocious winter gales and never landed, and though it was known that smaller, simultaneous French raids had been devised as vanguards to foment localised agitations and disturbances, and/or create diversions on Britain's mainland, the timing and purpose of Tate's desperate attempt was totally baffling. Its methods, however, are what specifically concern us.

Hoche's instructions to Tate were found on his person and made for such chilling reading they were promptly published by the London government to expose to the 'people of England' how abhorrent French intentions truly were, aiming as they did to 'disunite' and 'destroy' *all* ranks of English society.[9] Logistical matters ranged from disarming and subverting the militia to fording rivers, burning arsenals to opening prisons, and enticing the dispossessed to plunder. The orders were infused with virulent ideology justifying how Tate was to land his men anywhere remote but practicable, spread 'panic', and 'strike terror and amazement' by sabotaging naval and port infrastructure. Even Tone, who as a bilingual staff officer had been tasked by Hoche to translate them into English, despite being an ardent revolutionary and able propagandist himself, was painfully conscious of 'what misery the execution of the orders' entailed. He noted in his diary that his commission had done little for his 'morality', being forced to transcribe how Tate was to take Bristol, 'burn it to the ground' or produce its '*total* ruin', reducing thousands of families to beggary.[10]

Why land in Wales when the strategic objective was Bristol? There are evidently contradictions in these French plans. The orders were originally devised for a landing at Cardigan Bay, either from there to cross over to Ireland, or to act as a diversion to a major expedition to Ireland in November 1796. These plans were then amended to include a raid of Bristol. By February 1797 the landing on the British mainland was to be merely a raid: Tate was to land somewhere poorly defended with only *c*.1220 men, proceed inland, by night, march on Bristol to destroy its port infrastructure, anything linked to trade, commerce; then proceed inland to Chester and Liverpool to wreak further destruction.[11]

The type of mission entrusted to Tate was specifically called '*chouannerie*' by the handful of men who planned this in the hope it would engender an uprising in Wales and Ireland as a part of French military strategy against Britain, *c*.1795–1798. The concept evidently derived from the various savage tactics of the Chouan rebels in the recent counter-revolutionary insurrection in the west of France explored by Alan Forrest in his contribution to this volume, under the leadership of Hoche who had commanded republican troops in the Vendée. These planned raids on the British Isles (Tate's was the only one to succeed in coming ashore) were to be carried out by units (often free corps) of the French state army as irregular, small-war type operations. Motivated largely by the quest for revenge for English support of counter-revolutionaries and their 'vomiting émigrés, arms and gold on to [French] coasts', this reflected the Anglophobia of some of the planners. In wanting to incite an uprising through such an incursion, this could be seen as a precedent for CIA operations from Albania in 1949 to the Bay of Pigs invasion of 1961, or of focoism à la Fidel Castro and Che Guevara.[12] Often expressed in paroxysmic language, the plans fused a revolutionary ideological subtext with technical-tactical provisions and are insightful, even if only two were implemented, albeit unsuccessfully. Their overlapping goals included fomenting insurrections on the enemy's home front, but also using dirty tactics to disrupt domestic security, thus dragging civilians into the mayhem in an overlooked chapter of the total war of the age.

Via Ireland to Britain

Ireland's geostrategic position and vulnerability had long fascinated Spanish and later French naval planners and buccaneers, as both a base to invade Britain and gateway to the Atlantic and Caribbean. Kinsale harbour is only *c*.260 nautical miles from the French naval port of Brest (Brittany) and a single day's march to Cork, at this time the European port 'nearest' to the Caribbean.[13] In 1760 the privateer Thurot had briefly raided the north-eastern fortified town of Carrickfergus with *c*.1000 men, exposing Ireland's weak coastal defences.

The main goal was not to attempt anything militarily, but to divert attention from planned larger-scale assaults on mainland Britain and so distract domestic security, a strategy which would partially underpin the later plans which concern us. As war shifted towards ideological struggles between peoples and nations, new political motivations inspired French military strategy, and the idea of exploiting the growing disaffection in Ireland by stirring secret societies was also mooted (if never fully acted on). By 1767, one plan even proposed that if England committed a hostile act against France, Ireland should be seized and a republic established there under French protection.[14] During the American War of Independence a common view was that the 'face of war had completely changed', but that it was the English who had changed it by their aggressive descents on France's Atlantic coast during the Seven Years War.[15] One Patrick Wall suggested that Irish Catholics, who made up two-thirds of that 'bellicose nation', could not remain much longer in their current 'violent state of slavery and misery'. The American Revolution had inspired secret rural societies, but they could not act without arms or money. Wall was certain that they would vigorously support the first force arriving to free them. Yet after recalling the poor state of Ireland's defences, Wall asserted that no expedition of less than 20,000 troops could succeed. He put forward another interesting argument: A foreign army landed in Ireland could also deprive England of the substantial numbers of Irishmen enrolled (or rather, press-ganged) into the Royal Navy, and it was claimed nearly two-thirds of the men below deck were Irish (probably an inflated figure).[16] Overall, pre-revolutionary plans agreed that for logistical reasons (but not exclusively), only smaller raids manned by expendable forces were feasible and the endemic problems which the navy faced during the 1790s would only exacerbate this. However, French revolutionary zeal was to transform such invasion plans into liberating crusades.

A legendary, ambiguously worded, translated, and widely circulated French decree of 19 November 1792 promised to grant 'fraternity and assistance to every people who wish to recover their liberty'. As the American ambassador Alexander Hamilton reported to Washington, it did indeed 'excite fermentation' among British and Irish reformers, and a few radicals and exiles began networking in Paris to obtain France's intervention.[17] In February 1793, France's declaration of war on Britain sparked enthusiasm for an invasion of the British Isles, though no plans would materialise until June 1796 and Hoche's appointment to 'detach Ireland from England' and thus diminish her dominance of the seas.[18]

From the outset, French policy would be weakened by the lack of reliable intelligence on the actual state of disaffection, and degree of military preparedness, of potential Irish rebels, but one early and presumably reliable source was the former consul and attaché in Dublin, Charles Coquebert de Montbret. He had, however, somewhat overstated the claim to Lebrun, the French Republic's first Minister for External Relations, that the 'ferment in Ireland was most certainly the precursor of a revolution' which the inevitable war between

France and Britain would precipitate.[19] Furthermore, Irish seamen in the British navy (praised by Coquebert as 'intrepid corsairs', ideal for irregular operations) could help block ports and stifle commerce by enticing the enemy's forces to desert. Some 4000 or 5000 men debarking in the north-west with light artillery and a well-worded manifesto might form a rallying core for the 'friends of liberty'. Ireland's north-west – especially Ulster – was the most densely populated, politicised, and radicalised region, even if this polarised only later in the decade into sectarianism. At the time of the French Revolution, the hope was still that the core of an Irish rebel army would combine both Protestants and Catholics.[20]

The manifesto itself would be a vital propaganda exercise outlining the political aims of the invaders, which would also rally volunteers to join them. Thus psychological warfare could support military operations. Lebrun privately assured one exiled Irish officer that if despotism were overthrown in his original fatherland, it would be the French nation's duty to assist and *make any attempts in her power* to ensure liberty and equality would reign there'. No overt military promises were made, as France wanted to be able to deny that it was inciting insurrections and disturbances in neutral countries. Ideally, the Irish would rise up first.[21]

Lebrun was also liaising with his eccentric but enterprising chief translator, Nicholas Madgett, whose broad job specifications also encompassed communications, propaganda, improvising intelligence gathering, advising on policy, and now war. Madgett, another Irish exile, suggested that the war ministry approve missions to Britain by trustworthy Irish patriots who could proselytise in the press and political gatherings to 'spread the principles of liberty'. He also secured money for the translation and printing of seditious handbills targeting English sailors.[22] His nephew Sullivan worked under him and had helped smuggle patriotic tracts into England, a task facilitated by his official presence in France's Atlantic seaports as an inspector-interpreter for British prisoners of war. On one 'special mission', 200 prisoners had 'offered' to serve in the Republic's navy due to his 'zeal in preaching the principles of our revolution to the prisoners of war... and contempt and horror for King George and his ministers'.[23] Madgett and Sullivan were valuable local informants to Tone and continued to be involved in this recruitment practice to help man the French expeditions to the British Isles in 1796.

Three plans for an invasion

Three overlapping plans for small-scale secret expeditions to both mainland Britain and Ireland, undated but written around February to April 1796, all talk about creating a *'chouannerie'*.[24] Their scope is best understood by first looking at valuable insights from Tone's negotiations from February to April 1796, which broaden our understanding of the evolution of this strategy and how Ireland came to be implicated in it. Barely a week after launching his lobbying campaign, Tone was assured by Madgett that the French were quite serious about an expedition to Ireland; but, would the Irish rise spontaneously before a French landing? Tone was adamant they would *not*, and was dismayed when told that no large fleet could be committed given the poor state of the French navy, and only 2000 men spared. But someone would be sent to sound out the Irish among the British prisoners of war and cajole them into joining the expedition, then have them promptly exchanged for French prisoners. Totally disagreeing with this type of recruitment and scale of plan, Tone answered that only an 'imposing force' of troops, mostly artillery and who had seen hard service, led by the best general France could spare, would prevent 'an immense effusion of blood'. It could also train and discipline what would become the Irish army, confront the opposition, rally the people, and entice the militia to desert and join the invaders.[25] If the Irish insurgents did not first gain some advantage in the field and seize Dublin, they would have to 'go close hauled', by which Tone meant small, furtive operations. Even with 5000 French, the insurgents would be forced to proceed on a 'revolutionary plan' relying on 'the sans-culottes of Ireland', by which he probably meant untrained, inexperienced, and poorly equipped fighters, if zealous patriots, drawn from the poorest of the poor. It meant forcibly requisitioning every man, horse, guinea, and potato in Ireland, yet being able to achieve little militarily.

Tone's musings also echoed conversations around him about fitting strategy – fuelled by the prevailing Anglophobic mindset – to the means available. Anticipating his commission to serve in the expedition, he was avidly reading military manuals and regulations, attentive to ongoing debates about reform and rationalisation of the forces, but also to cultures of combat. One rambling soliloquy on military discipline and repression of drunkenness drifted into a positively Saxean *rêverie* on the uselessness of distance firing updated for a nation in arms, culminating in this portrayal of the new Irish freedom fighter: 'The *arme blanche* [naked blade] is the system for the French and I believe for the Irish... for poor Pat is very furious and savage, and the tactics of every nation ought to be adapted to the national character.'[26] As the Irish supposedly had more 'animal spirits' than the English or Germans, he voted 'for the bayonet', and wished to instil into them 'the character of the French soldiery'. He even affirmed 'surely we [the Irish] can do as much as the *Chouans*, or the people of the Vendée!' Yet he vehemently opposed the double-edged French strategy taking shape for Ireland.

By April, General Henri Clarke, the French-born son of an exiled Irish officer, now head of the Directory's military topographical bureau (later minister of war in 1807) gave his support to the project of turning Ireland into 'Britain's Vendée'.[27] He asked Tone to write up a 'short plan for a system of *chouannerie* in Ireland', because the government wanted the landing of 'a parcel of renegades' to distress, distract, and embarrass the authorities to precede any serious attempt.[28]

Dismayed, Tone in a brief, sharp memo boldly urged the formulation of clear military objectives.[29] He felt that the French Republic should invade Ireland with 'more enlarged views and a *sounder policy*' (my emphasis). 'Chouannising' Ireland would inevitably lead to 'indiscriminate plunder', and only cause easily and harshly suppressed localised risings. The English militia would be shipped in with fearsome irregulars to replace the Irish one, an official tactic to 'awe' both countries; hastily trained and inexperienced, unsuitable for police work but far removed from home, they would have less trouble acting ruthlessly. The Irish people's minds were prepared enough and needed no rousing or stirring. If, as hoped, this *chouannerie* did trigger insurrections, 'the most ardent... and bravest' Irish peasants would be sacrificed first. France had been provoked by English intrigues on French soil, Tone acknowledged, and had 'every right... to retaliate on England the horrors... and abominations... of the Vendée'. Ireland had never concurred willingly in any measure to distress France, yet this plan would 'make her suffer for the crimes of her oppressor'. Finally, on a tactical note, he signalled that Ireland was not a country adapted to support a Chouan-style insurrection, being far less wooded than Brittany and the Vendée.

Though Clarke reassured Tone that the plan had been dropped, in reality three further memoranda were being circulated and discussed. By June 1796, when the Directory appointed Hoche as commander of a major invasion force for Ireland (amounting to nearly 50 ships, 14,000 troops, and substantial supplies of munitions), plans for smaller raids in Britain had been fine-tuned and were now firmly embedded into the overall strategy.

Hoche had initially called for an offensive war on Britain in October 1793. Envisaging a major commitment in terms of men and matériel (40,000 troops), and the slightly utopian vision of a fleet of armed merchant vessels bridging the channel to transport them, he claimed that 'No manoeuvres or art: steel, fire and patriotism' were all that was needed. 'What rules of war are we expected to follow with Barbarians who fight us with poison, assassination and arson?', he asked, wishing to be the first to set foot on the soil of these 'political brigands'.[30] Another plan called for nothing less than the devastation of London and destruction of the arsenals; the petitioner stated this required a substantial force (a totally unrealistic 100,000), and executed as a bold, sharp, and rapid strike.[31] Hoche was galvanised when he learned of British conspiring with Vendean leaders, and in March 1795, weeks before the British landing in Quiberon, he launched the Committee of Public Safety to deploy a pre-emptive strike against England. A year later, the Directory wrote to Hoche tasking him with promptly and secretly organising a 'chouannerie of sorts' to foment agitation in England (and in turn the long-awaited revolution in Ireland), passing on some thoughts on the matter by General la Barollière.[32]

La Barollière's plan

La Barollière's note to Clarke, which only refers to England, hints that he had been asked to contribute his own views.[33] Given his personal trajectory, it is surprising that la Barollière's formulation of counter-chouannising abroad has received so little attention. His career spanned the transition from royal to republican service, and in 1791 he was promoted to lieutenant colonel in the Ninth Regiment of Mounted Chasseurs (1791), then to divisional general in 1793. He then commanded republican troops in the Vendée and was still active there in April 1796, around which time this note was presumably written. La Barollière was cautious about expecting a revolution just by 'throwing' folk into England, though it might trigger one at a later date. Then:

> If, as hoped, by landing [French] troops in England, regardless of numbers, the English would be forced to provide escorts for their public transports, this would be very advantageous. In the western army of France, more than 60,000 men were employed merely as escorts, yet despite these impressive numbers, the government was experiencing losses daily, which demonstrates the numerous advantages of the chouannerie plan, *despite its immorality* [my emphasis].

The escorts would be needed because la Barollière also advocated that all public conveyances be attacked and looted, but such police duties would divert the escorts from more effective combat use elsewhere. He also had views on direct action aimed at *fomenting* disturbances, if not full insurrections in the political sense:

> Therefore it appears one would instruct the commander put in charge of the chouannerie in England to follow the same principles as the *chouans* did in France, to engage in combat the least possible and only when they are absolutely forced to, or when they [outnumber their opponents] six to one.

The following recommendations opened with the iconic incantation dating back to the foundation of the French Republic in 1792 followed by more prosaic aims:

> To announce war to the castles and peace to the cottages; to declare to the troops that anything they seize in England is legal, and theirs to keep;
> To open all the prisons; arm all detainees to increase the force; burn without restraint anything linked to the marine...
> To have not the slightest pretention of [being] an army corps, supporting a said party, but to renew what the filibusters used to do in the islands [of the Caribbean]. That each individual who will go play the chouan in England will do so with the plan to there amass or steal 100,000 francs, to then end his career in comfort.

The released convicts thus shipped to Britain would be assured that they would receive 'formal absolution' from the government after a stated period of time, and be allowed to retain their 'gains'. French deserters recruited in England, be they émigrés or released prisoners, would also be pardoned for services rendered. To achieve the intended goal, la Barollière advised on the logistics of splitting the contingent into smaller units leaving separately from different French ports, as it would be difficult to keep the destination secret when fitting out and loading a ship for even 1500 to 2000 men. Landing on several parts of the coast made sense, as it would be smaller corps of troops who would 'spread desolation to a wider expanse of the country'. The 'leaders' were to be 'intelligent', carefully chosen, issued with detailed instructions, maps, and inland rallying points; munitions were the only essential item to bring.

Carnot's plan

The subtext of the most extreme elaborate and effusive of the three plans, that of Lazare Carnot, one of the revolutionary ministers of war, clearly advocates the use of terror as the means to an end. This was, again, to embarrass the English government and so hand 'back some of the horrors of the chouannerie which they had organised and fed in the heart of the Republic'.[34] The force would be manned by men full of audacity, fearless when faced with danger, and capable (like the aforementioned filibusters) of carrying 'fear and death in the enemy's midst'. Some might be released convicts with the physical and moral 'qualities' needed for such an operation, and even court-martialled soldiers condemned to wearing shackles. Lured by booty, they would be led to believe they would be pardoned and live out their days enjoying the fruits of their easily snatched fortune in some distant colony, 'like Cayenne' (a tropical hellhole). Carnot signalled that such a smallish expedition of only 2000 men could only succeed with 'violent and extraordinary' measures. Prisons would be opened and workers, indigents, and malcontents would be enticed to join.

To fulfil its objective, this 'war' must be conducted 'in the same manner as the one waged on us by the *Chouans* in the Republic', the defining reference now embedded into semi-official policy. Carnot was a military engineer by training but who valued the art of skirmish, and so launched into a frenetic enumeration of its methods, which included multiple attacks on posts spread over a wide area, to convince the enemy of numerical advantage. As in traditional small-war operations flanking regular armies, the enemy was to be harassed and made to cover a lot of territory in a single day, echoing Sun Tzu whom we can thus surmise Carnot had read in the recently published translation into French:[35]

> fall on him when he is weakest; disappear beyond his reach when he is strongest, return and charge, attack even by bayonet, taking cover behind hedges and in ditches, preferring to act in the rain or fog; pursue him to the end, when he is beaten; retreat a lot, easy in [sectioned] terrain; destroy bridges... block roads, cut off communications and even stop and plunder public transport and coffers... destroy anything belonging to the navy, seize government funds, summons resisting communities to hand up their arms, exact levies from them and militarily deal with any opposition.[36]

Humbert's plan

General Jean-Joseph-Amable Humbert's 'Ideas for a chouannerie in Ireland' mirror the two previous ones. Humbert, a close associate of Hoche's, has been much maligned by history, denigrated as a near-illiterate brute suited for this type of operation, yet his plan is pragmatic, more militarily 'tactical-technical' in its views on success than driven by ideological fanaticism.[37] His initial mission was, under Hoche's supervision, to lead an ambiguously named Legion of Francs of c.1000–1050 men on a raid of the (appropriately) wooded region of Cornwall, there inveigle tin miners to rise up in arms, then make his way to Wales. If the Welsh were not prepared for an insurgency (French intelligence on the situation in Wales was inadequate), the invasion party could simply wreak havoc, destroying bridges on the way. Rather than emulating the intrepid Franks of old, his men would include foreign deserters gathered in expendable free corps to be 'thrown onto the coasts of England'.[38] This mission, subsumed with Tate's as a diversion into Hoche's major expedition to Bantry, never sailed. Nevertheless, as Humbert was the only French commander to actually harass the British enemy at home during his brief occupation of Ireland (August–September 1798), his views are most insightful. They also reflect his own experience of the Vendée, where he not only saw combat but had successfully negotiated with the enemy, at Hoche's command.

In 22 clear and concise articles, he outlined in mostly military terms what was expected of the reliable, audacious, and robust troops. Leadership skills were not overlooked, as officers should be 'intelligent', literate, and totally devoted to the government.[39] Fitted out in grey or brown, i.e. camouflage, the men were to seize the most advantageous positions, elevations or wooded areas. Dispersing into cantons and companies to facilitate living off the land and locals, unit leaders were to enforce the strictest discipline and harshly punish drunkenness, but also monitor moodiness as despondency was noxious to courage. They were to be honest and restrained in order to 'make partisans', rally locals and possibly train them, and 'employ political means' to that effect, presumably engage in republican proselytising. Local spies were to be recruited even among local women, children, and the elderly, and if funds were needed for this requisitioning was acceptable but without unduly antagonising the population. Yet he recommended using 'vigorous means' so as to not 'diminish activity', so vital in 'this type of war'. Carnot had stated that leaders had to display a quasi-despotic power to be effective – again, one is reminded of the tactics of focismo in the mid twentieth century.[40]

Humbert's articles on combat tactics (14 and 15) mirror Carnot's plan. Seizing enemy horses, the skirmishers were to 'pursue' the enemy 'to extinction'. Leaders were to remind their men that it was the English who had organised 'all the civil wars in France' and consequently 'we' [the French] had to seek vengeance. As *only* senior officers could instigate 'hostilities', Humbert favoured a fairly hierarchical, authoritarian form of military enterprise. Echoing general calls not to alienate potential allies, he insisted that leaders were to impose the respect of persons and property and 'exercise the greatest impartiality'. Unit leaders were only to act once the inhabitants had declared their readiness to serve them, but if they did not act as partisans or were 'in rebellion', exactions would be imposed and their arms seized to repress them, thus instilling fear. Humbert was never graphic when hinting at violence, but did outline the need for coercion to effectively 'submit' the country to the invading force's overall military control.

His own ability to do so was tested to the limit in the very complex setting of the west of Ireland in 1798, as we shall see.

Hoche's instructions for Tate's invasion

Tone had described Tate's mission – the invasion of Wales and south-west England – as a 'buccaneering party', speculating that if he was 'a dashing fellow with military talents', he'd 'play the devil in England before being caught'.[41] Hoche had sent him to a prisoner-of-war camp at Pontanezen (north-east of Brest, Brittany) to persuade the men to 'serve aboard the French fleet', while 'never' mentioning the destination. Therefore most did not suspect that they would be taken back on their native country where they might be exchanged for French prisoners; only, as native subjects of the realm they would probably be charged with high treason for conspiring with the enemy.

Tone's recruits were mostly Irish, and though a democrat and lawyer by training, he did not comment that he had found the men 'half naked and half starved', and had plied them with wine and food before they were marched off. During interrogations of Tate's men after his surrender, the testimonies of those who had been prisoners of war in France were scrutinised by a House of Commons inquest into the conditions of their detention.[42] The subsequent report stated that some had been 'taken advantage' of when intoxicated, 'all efforts used to inveigle them' into joining the French service, and several hundred had been successfully 'debauched into the scheme'. Pontanezen was cited for its lamentable conditions, but no reference was made to any responsibility of eloquent Irishmen for the debauchery.

In his orders to Tate, Hoche never referred to the chouannerie as a model though, but his personal input can be summarised as follows. He ordered the *total* destruction of Bristol by setting it ablaze at night. The main goals were to spark off a Welsh uprising, if possible; to interrupt and hinder the enemy's commerce; to prepare and facilitate an invasion of Ireland by distracting the English government in south-west England and Wales. In view of inadequate intelligence, the French had totally miscalculated Welsh disaffection.[43] They hoped that in addition to the desperate poor, they could lure vagabonds, idlers, and malefactors into these companies of insurgents with money and drink. These were to lead the uprising. To incite the poor to plunder, their natural envy of the rich would be encouraged by 'inveighing against the government' as the cause for their distress; proselytising would presumably be left up to the English-speakers among Tate's men. The destruction of bridges and communications would also, Hoche noted, impede the army's movements and supplies, but he specified that even 'private carriages' were to be plundered. The enemy's troops were to be 'seduced' to desert and the militia disarmed, because – an echo of Carnot – such 'extremities were rendered necessary by the example set by our cruel enemy'. Generally avoiding direct encounters with regular forces, if such an engagement proved inevitable (even against an enemy twice the size), Hoch admonished Tate to remember that he was 'now a Frenchman, inasmuch as you command Frenchmen, and let that incite you to attempt a brilliant stroke'.

French Revolutionaries often asserted that France had no quarrel with the people of England, groaning as they were under the yoke of tyranny. Nevertheless, the type of war Hoche envisaged was 'terrible'.[44] The invading forces were to split into columns to 'spread the panic... generally'. Property belonging to 'men of rank and high fortune', the military, and militia was to be plundered in 'predatory excursions' by detachments of 300–400 men. Troops were to land with munitions only, the inhabitants, especially the gentry, were to 'supply their wants'; if a town or village did not comply, it would be pillaged; if locals did not warn of the enemy's approach in an area where a column was posted, they would be 'given up to fire and sword'. At the same time, there seemed to be (quite unfounded) hopes that urban populations including in Cheshire and in Liverpool would rise up in sympathy with the French.[45]

Humbert's campaign in Ireland 1798

Though British fears of a French invasion peaked for the last time *c*.1803–1805, General Napoleon Bonaparte had famously concluded by 1798 that to attempt such an operation 'without being master of the sea would be the boldest and most difficult ever attempted'.[46] In contemplating an invasion of Britain, a plan he wisely abandoned, he never quite showed Carnot's fanaticism.

After surrendering, Tate and his officers signed a statement for their superiors in France citing the insubordination of their men and the total lack of subsistence as key factors in their inevitable decision.[47] Tate had offered his services to Clarke to arm the discontented of Jamaica and 'harass the [British] enemy by desultory operations of plundering parties' in Bermuda and entirely 'without expense' to the French Republic.[48] In all the plans discussed, it was implied that such deployments would be self-reliant detachments, in effect marooned as the naval transports would sail back to France after they disembarked. While it was totally impracticable to envisage French management of an occupied zone across a body of water (as in a 'liberated' Ireland), Tate operated as a sort of private entrepreneur but also a legitimate belligerent with a commission in a national army. This is not the only paradox or contradiction emerging from this episode, which in many ways fit the historical pattern of fifteenth- and sixteenth-century condottieri, operations described by Benjamin Deruelle in his contribution to this special issue, or the operations of a Drake and Raleigh, minus the queen's letters of marque to give them legitimacy: the planners clearly had envisaged their menace as 'a silent threat', their men 'not disclosing' themselves until the moment of attacking.[49]

A handful of men devised these plans and were creating a new dogma of, as Hoche had put it, a terrible form of warfare. Yet when summoned to lead a chouannerie abroad he had sarcastically reminded the Directory that he had previously 'been told *we should contain [comprimer] terrorism*' (his emphasis).[50] We note la Barollière's own admission of 'immorality': 'speak much of liberty, but have absolutely only one plan: to destroy, and not edify'.[51] Tone had discovered that competition within France's executive was hampering decision-making. In any case, Carnot's secret promotion of his pet scheme and the employment of the 'scum' taken from prisons to man the expeditions was heartily disapproved of.[52] Even to contemporaries, these schemes were 'absolutely extraordinary' in a negative sense, and the expeditions were more aptly called 'brigandage' than military operations.[53]

Yet, the conduct of Humbert and his officers in Ireland proves that they saw it as their duty to uphold prevailing views on the laws of war in terms of their interaction with civilians, rules which Tate's mission brutally breached. Humbert's brief campaign was a political failure as he landed in August 1798, two months after the long-awaited Irish rising had occurred but had been comprehensively quashed. (Indeed, the British counter-insurgency campaign in Ireland of 1798 is described by Lawrence James as 'one of the nastiest wars ever fought in the British isles'.[54]) Humbert's few military encounters with the combined Crown forces were symmetrical affairs, but relevant to this discussion is the strict observance of respect for property and persons by the French left to manage the occupied zone (in County Mayo). The French had been led to believe that Irish rebels would flock to join them, but they did not. Instead, the war of liberation was bogged down and the French as occupiers felt compelled to assume local policing and peacekeeping duties. In a tense setting which was far from binary, their arrival had exacerbated local tensions and triggered plundering raids by destitute peasants on affluent loyalist Protestants, who later praised the French for duly protecting them. Colonel Charost, commander of the seat of occupation (Killala) had even made it clear to locals that he would have to hand in their plunder: he was, as he told them, a leader of a brigade, not of brigands. After his defeat and surrender, Humbert and his officers were treated as legitimate adversaries: they were ceremoniously escorted to Dublin where they were wined and dined. By contrast, the Irish rebels left on the battlefield of Ballinamuck (County Leitrim) were treated as illegitimate rebels, and pursued to extinction by the cavalry, much as the Scottish Jacobites had been after the 1715 and the 1745–1746 insurgencies.

Also significant were non combat-related aspects of operations such as propaganda and psychological warfare, in which demonisation of the enemy was crucial, here in justifying revenge. Tone had certainly experienced deep misgivings about his own hand 'in framing' Tate's orders, but had laid all the blame on the much detested English who were keeping his 'country in slavery', and concurred with the French impulse to thus 'avenge a wrong'.[55] Despite his occasionally naive views on revolutionary warfare, he was hardly the first freedom fighter to romanticise his mission, and also produce some sharp observations on the vagaries of strategic planning. He had also accurately predicted the draconian measures employed to disarm the Irish rebels, crush their rebellion, and break the spirit of the people. The route of the irregulars shipped in for this purpose could be traced by the smoke and flames of burning cabins and trail of dead bodies. If these resembled the infernal columns that the French state army had deployed to defeat the counter-revolution in the Vendée and in Brittany, it was because they too had become the model for a new form of small war, as one Anglo-Irish commander had told his superior:

> I look upon Ulster to be a *La Vendée* [*sic*] . . . . It will not be brought into subjection but by the means adopted by the [French] republicans in power – namely spreading devastation through the most disaffected parts.... Laws however strict will not do, severe military execution alone will retrieve the arms from the hands of the rebels.[56]

Acknowledgements

The author would like to thank Professor Alan Forrest for his helpful comments and suggestions.

Notes

1. See Stuart Jones, *The Last Invasion of Britain* and Rose, 'The French at Fishguard'. Numbers vary, but the safest estimate is that Tate had *c*.1220 men, though his initial orders refer to 1050: [Hoche], *Instructions to Tate*, v. Hoche's original has never been located in French archives.
2. Service historique de la Défense, Vincennes [henceforth SHD] SHD/Army/17y^d 12 Gp: William Tate (personal file).
3. [Tone], *Writings*, 2: 370.
4. As a brigade colonel, then promoted adjutant-general in October 1796, under the pseudonym James Smith: SHD/Army 17y^d 14 Gp and Kleinman, 'Brave de plus' . Tone was a founding member of the Society of United Irishmen, set up in Belfast and Dublin in 1791, initially to reform domestic politics though some had more militant views. Banned in 1794, they reformed and expanded as a clandestine revolutionary body, were the main interface with France and organisers of the rebellion in May–June 1798.
5. I am grateful to Dr Paul Rich for this information.
6. Gastey, 'L'étonnante aventure'.
7. The most thorough study of the Franco-Irish strategic alliance remains Elliott, *Partners in Revolution*; for the chouannerie plans, see 77–134.
8. See contribution by Alan Forrest elsewhere in this issue.
9. [Hoche], *Instructions to Tate*, iv, and following sentence, 12 & 8; Jones, *Last Invasion*, 62–5. The terms 'Great Britain' or 'British' rarely, if ever, appear in French contemporary sources consulted for this article; I have literally translated their use of 'England' and 'English' to designate the state France was at war with, except when the purely geographical term (e.g. mainland Britain) is more relevant.
10. Tone, *Writings*, 2: 399, my emphasis. For the strategic role of military linguists, see Kleinman, 'Translation, the French Language and the United Irishmen', summarised in 'Amidst Clamour and Confusion'.
11. Tone, *Writings*, 2: 399.
12. Hoche to the Directory, 28 April 1796, in Guillon, *La France et l'Irlande*, 89. For CIA tactics, see Heuser, 'Covert Action'.
13. See also García Hernán et al., *Irlanda y la Monarquía hispánica*.
14. 'Needs and means to diminish the forces of England' Colonel Edmund O'Reilly to Choiseul, 7 October 1767, cited in Beresford, 'Ireland in French Strategy', 286.

Beresford discusses both strategy and political motivation behind various plans, many commissioned from, or proposed by Irish officers serving in France. However, until the 1790s Irish disaffection did not significantly encompass separatism.

15. Beresford. 'Ireland in French Strategy', 420, and following, 424, citing [Patrick?] Wall's memorandum of 4 March 1780, SHD/A/M&R 1418, 281–305. Many of these plans were collated in 1796 by General Henri Clarke and read with interest by both Hoche in 1796, then Napoleon Bonaparte in 1797–8 and 1804–5: e.g. Archives nationales de France [henceforth AN] AN/AF III 186^b/857–60 and AF/IV 1671.

16. Restrictions on Catholics bearing arms had been lifted by 1794, but minimal numbers chose the French over the British army.

17. Hamilton to Washington, 'Answers...' May 2 1793, emphasis as per original: in Hamilton, 'Hamilton to Washington'.

18. SHD/Army/11B^1, Directory to Hoche, 19 June 1796.

19. Archives diplomatiques de France [henceforth AD], AD/CPA/584, fols.214–217^v, Coquebert to [Pierre-Henri-Hélène-Marie Tondu, called] Lebrun, Paris, 18 December 1792, and following.

20. Elliott, *Partners*, 20–1, 125, and following.

21. AD/CPA/587, 9$^{r.-v.}$, Lebrun to Lt. Colonel André MacDonagh, 1 March 1793, my emphasis.

22. AD/CPA/587, 9$^{r.-v.}$, Madgett to Lebrun, 13 March 1793, fol. 20^r, and 46^r. Secret expenditure records confirm that sums were allocated on April 26 1793 for the translation of placards addressed to the 'brave English seamen': AN/AF/II, 7^r. See also AD/CPA/588, 480–1, undated but *c.* late May 1794, 'Address to the people of Ireland by citizen Madgett Head of the Translation Bureau before the Committee of Public Safety... the purpose of which is to waken the hatred of the Irish against their eternal English oppressor'. The true extent of the role of Irish agitators in France, namely one Wiliam Duckett (also a translator/propagandist), in the British sailor's mutinies at Spithead and the Nore, April-May 1797, merits a dedicated study: see e.g. AD/Personnel/1/25 [Duckett].

23. AD/Personnel 1/65, 58^v, Sullivan to Minister for External Relations Delacroix, 30 October 1795. Sullivan served as a captain and bilingual adc under Humbert in Ireland and took part in haranguing locals to join.

24. The three main plans, undated but catalogued with material related to plans for the Irish expedition circa April to June 1796 in SHD/Army/11B^1 are attributed to generals La Barollière, Carnot (a serving Director, i.e. member of the executive government), and Humbert, see notes 30, 31, and 32. See also Come, 'French Threats to British Shores'.

25. Tone, *Writings*, 2: 70–1, and following.

26. Ibid., 2: 138, and following, 72.

27. Elliott, *Partners*, 85–7.

28. As note 20: 2 April 1796, 140, and following; emphasis as per original.

29. AN/AF IV/1671, 65^r-67^v, 'Memorandum to General Henri Jacques Guillaume Clarke on the encouragement of *chouannerie* in Ireland, 4 April 1796' [in English], reprinted in Tone, *Writings*, 2: 144–5, and following, 174.

30. Hoche to the Committee of Public Safety, 1 October 1793, in Guillon, *France and Ireland*, 73.

31. Gauthier, a naval official, in Debrière, *1793–1805*, 1: 30.

32. Directory to Hoche, 18 April 1796 in Debidour, *Recueil*, 2: 176–7.

33. SHD/Army: Jacques Marguerite Pilotte de la Barollière (1746–1827): GDI 7Y^d 33; 11B^1, 'Note for Clarcke', also in Débrière, *1793–1805*, 1: 64–6, and following.

34. SHD/Army/11B[1]: [Carnot], 'Instruction [sic] to establish chouannerie in England', also in Savary. *Guerre des Vendéens*, 6: 333–7ff. My thanks to Alan Forrest for his insights on this section.
35. Sun Tzu was first translated into a European language by Amiot, *Art militaire des Chinois*, which includes 'Les Treize articles sur l'art militaire, ouvrage composé en chinois par Suntse...'
36. As note 31.
37. SHD/Army/11B[1]; Jean-Joseph Humbert (1767–1823) personal file: 482 GB 84[d] 2e série; Ideas to establish a chouannerie in Ireland' [Idées pour établir une chouannerie en Irlande] in Debrière, *1793–1805*, 1: 61–4, and following.
38. Directory to General Beurnonville, 18 April 1796, in Debidour, *Recueil*, 176, outlining the plans to assemble foreign deserters in Atlantic ports as 'free corps', i.e. irregulars; we note the ambiguity as in French 'franc' can mean free, or Frankish (but also frank as in English).
39. Humbert. 'Ideas' as note 32, and following.
40. As note 31.
41. Tone, *Writings*, 2: 397, 371, and following.
42. House of Commons, *Report*, 9. Conditions of detention in France were under investigation by Britain, and a [bilingual] cartel for pow exchanges and detention standards was agreed in London on 13 September, 1798: SHD/Marine FF[1]/33/V1.
43. Jones, *Last Invasion*, 63, 205–11. This is reminiscent of Che Guevara's lack of intelligence on rural Bolivia which led him to miscalculate his moves there two centuries later.
44. Hoche to the Directory 9 June 1796, in Guillon. *France and Ireland,* 90.
45. [Hoche], *Instructions to Tate*, i.
46. Bonaparte to the Directory, 23 February 1798, in Bonaparte, *Correspondance*, 3: nr.2419, 644.
47. Jones, *Last Invasion*, 278. All were exchanged, even the Irishmen, as part of the ongoing Franco-British talks.
48. AN/AF/III/ 186[b]/858/62[r], Tate to Clarke, 27 July 1796.
49. Lawrence, *Seven Pillars of Wisdom*, 200.
50. Hoche to Directory 28 April 1796, in Guillon, *France and Ireland,* 88, emphasis as per original.
51. 'Note for Clarcke', in Débrière, *1793–1805*, 1: 65.
52. Elliott, *Partners*, 86.
53. Debrière, *1793–1805*, 1: 61, 66.
54. James, *Warrior Race*, 256.
55. Tone, *Writings*, 2: 399; Heuser, *Evolution of Strategy*, 388.
56. General Thomas Knox to the Earl of Abercorn, 21 March 1797, cited in Bartlett, 'Defence, Counter-Insurgency and Rebellion', 270.

The insurgency of the Vendée

Alan Forrest

Department of History, University of York, UK

The insurrection in the Vendée combined open warfare with the methods of petite guerre, ambushing French republican soldiers and cutting their supply lines to Paris. These tactics, when combined with the hatreds generated by a civil war, go far to explain to the cruelty of the conflict in the west and the depth of the hatreds it engendered. In republican eyes the use of guerrilla tactics was unjust and illegitimate, and they denounced their adversaries as common criminals and brigands, portraying them as backward, superstitious, even as subhuman, and in the process justified the savage repression they unleashed against them.

The civil war in the Vendée has become a critical part of the memory of the West of France and is a powerful element in its identity: a region that has willingly assumed the mantle of a martyred province after one of the most savage campaigns of the Eighteenth Century. The war is remembered in the West as one of principle; a bitter struggle between peasants and Catholics on one side, republicans and atheists on the other. It was a war for God, fought with all the intensity of a religious crusade. In purely military terms, it was a mixture of different types of warfare, ranging from ambushes and isolated attacks on republican columns, to open battles when circumstances allowed, an insurgency that profited from familiarity with the terrain and recruitment based on feudal obligation to local barons, a system that provided the Grande Armée Royale with some 40,000 men, and Charette with a further 5,000 recruits in 1793, numbers far beyond the capacity of the Republic. If it had a disadvantage it lay in the indispensability of individual commanders and the constant likelihood of internal bickering and petty jealousies amongst them.[1]

It was also, of course, an insurrection by a single region of France against the authority of the government, of the Revolution, and of the French nation. This is how it was seen by republicans and by the National Guards sent to repress it. It is how Michelet recorded it for future generations: 'a people so strangely blinded and so bizarrely misled that they took up arms against the Revolution, their mother, against the security of the people, against themselves'.[2] It was, moreover, of astonishing tenacity, a civil war that blazed in 1793–1794 and then temporarily died down before rekindling under the Consulate in 1800, and again after the declaration of the Empire in 1804.[3] For the inhabitants of the region it remained what it had been for their ancestors: a principled struggle against an authoritarian and centralist regime, a struggle for their liberties and traditions, a combat fanned by their priests and inspired by a simple and profound Christian faith. But we can legitimately challenge this rather sanitised version of the region's history. The rebellion was fomented at the moment when the first great levy of troops was announced, the *levée des 300,000*, the first attempt by the Revolution to force the young into uniform and to compel local communities to part with their sons. This is not a matter of idealism or the principled response of a united population. And for a generation of historians like Paul Bois and Charles Tilly, the origins of the revolt are to be found in social and economic interests, in village conflicts over land and property.[4] There was not a single Vendean response to the Revolution, but several, often deeply factious, responses. There were republicans as well as counter-revolutionaries in the west, communities riven by internal squabbles, villages and families torn apart by the war.[5] Neighbour denounced neighbour, brother fought against brother as the Vendée slid into the throes of a hateful civil war.

The movement had its roots in various local uprisings during 1793, popular in character, and specialising in sudden attacks and ambushes against the soldiers sent to destroy them. This was a war fought in the *bocage*, in the tradition memorably described by Balzac in his fictionalised account of the Breton *Chouans*,[6] whom he re-created in the guise of virtuous savages, to be compared, in the words of Jean-Clément Martin, 'to Mohicans or Hottentots'.[7] But the Vendean war was not all like that. In 1793 and 1794 especially, it was fought between armies, when, in the campaign known to history as *la Virée de Galerne*, a 'massed army of tens of thousands of soldiers and their families criss-crossed Brittany, reaching as far as Grandville', before suffering a decisive defeat at the hands of the Republic.[8] At the peak of the movement there were as many as 70,000 men under arms, fighting in formal units, marching in disciplined columns, and subject to standard military discipline. This was scarcely a 'small war'. The Vendeans passed easily from one style of fighting to another; indeed, it was only when they were no longer capable of gaining advantage in open campaign that they resumed guerrilla tactics, regarding this as their only method of self-defence. In that respect, their tactics can be compared to those of regular armies, which had often fought a *guerre des partis* during the limited wars of the eighteenth century. As open engagements were rare, they sought to harm each other by cutting off supplies, starving the enemy, attacking convoys, and seizing outposts – in short, as Sandrine Picaud-Monnerat reminds us, by all the methods of *petite guerre*.[9]

For their opponents, this was a tactic that was indistinguishable from banditry or brigandage, and guerrilla manoeuvres were denounced as a form of criminality and an outrage to the peace of civilised peoples. So much was opaque and unclear. Who was a soldier and who a civilian in this shadowy form of warfare?

Who had the right to seek protection from the violence of war? For the republicans these were critical questions that called out for answers. As the republican general Turreau recalls in his memoirs, there was little distinction between soldiers and civilians. 'The rebels gained great benefits from the friendly relations they enjoyed with the inhabitants who had stayed behind in their homes. Too cowardly to take up arms with them, they nonetheless secretly favoured their cause; they played a part as spies; women, even children were loyal and intelligent agents who informed the rebel leaders of the slightest movements of the republican army.'[10] It was this confusion between civilians and soldiers, made worse by the almost universal hostility which the army encountered throughout the west, which, said Turreau, went far to explain the brutality of his troops' response. And brutality was never going to be enough. To resolve the problems of the west, he knew, involved more than repression; it required 'the re-education of a whole people, since fanaticism and servitude were the natural companions of ignorance'.[11]

The war in the west rapidly embraced all the qualities of a civil war, which tore apart the sinews of society. For republicans, and especially those living in the cities of the west, it was equated with the blind reaction of backward and superstitious peasants, bullied by their priests and landlords into rejecting the ideals of 1789. In Maine, for example, which the rebels crossed three times during the Virée de Galerne, local Jacobins fully shared the terrorist impulses of the sections of Paris.[12] They justified the use of repression as a simple measure of policing, arguing that the nation had a monopoly of armed power and that there was no place for independent militias. To adapt the phrase of Reinhart Koselleck, 'the state had staked its claim to be the only legitimate authority in all cases of civil war, refusing to recognise fratricidal struggles or ideological wars between partisans and opponents of the Revolution.'[13]

The violence of the reactions the war evoked, on the left as well as on the right of politics, can be explained in part by the brutality of the repression of the insurgents by the republican army, but also by the prevalence of a Vendean mythology that was created in the war and which flourished during the nineteenth century after the restoration of the Bourbons. This maintained that the Vendée was a region of a unique Christian piety, a region born out of repression, a martyred province whose people had suffered stoically in the name of Christ and king. If the French Revolution had its bicentenary celebrations in 1989, and the Republic in 1992, in the following year it was the turn of the west and the bicentenary of this cruellest of civil wars. It was a year to wallow in tales of Christian fidelity and of a guerrilla war steeped in the romantic countryside of the west, a year, in short, to play to the hyper-romantic imagery of Balzac and Victor Hugo and the imagery of a largely mythical Vendée of Christianity and chivalry.[14] The truth was always going to be more complex, however. Among the rebels were deserters and stragglers picked off from the republican army as it passed and hastily incorporated into its ranks, soldiers more motivated by the quest for booty than any Christian ideals, men capable of the most astounding acts of cruelty towards those who fell into their hands, and civilians who, sadistically and unashamedly, perpetrated the most extreme acts of savagery on the bodies of their prisoners. Often, too, the conflict reflected the tensions and animosities that divided rival villages, rival families, or clans, creating bitterly divided communities that were ready to collaborate with approaching armies and to denounce long-standing enemies in order to assuage their desire for vengeance. Elsewhere the conflict assumed the traditional character of a struggle between town and country over land, resources, or local power. Many of these communities were routinely divided into opposing factions, assuming the identities of blues or whites. It is here that the deepest rancour and the most shameless collaboration were often to be found, in divided communities which, in the name of principle, pursued their own factional interests and fought their own, purely local, civil wars.[15]

It is easy to see why the conflict quickly assumed such bitterness. In the early moments of the rebellion, Paris and the Convention seemed oblivious to the gravity of the situation they faced; the reports that came in from the western departments were imprecise about what was going on, reflecting a tense political atmosphere where Jacobins and Girondins treated one another with open distrust. Everything was reduced to one essential truth: that there had been a counter-revolutionary rebellion, the work of nobles, Catholics, and royalists, and by communes that had been, in Turreau's words, 'infected by the spirit of Roland'.[16] But when news arrived of the fall of Cholet and the defeat of General Marcé, Paris began to take note. The deputies sent on mission to the west expressed growing alarm that Paris had failed to provide sufficient manpower to repress a rebellion that seemed to be increasingly successful and whose leaders were seen as well tuned to the mood of local people. In their eyes the rebellion could only be explained by the character of the local population and their simple, superstitious naivety, especially in the presence of the clergy; three commissioners sent to the army in the west reported that 'it is because of ignorance that we have suffered the unhappy war in the Vendée', adding ominously that this same ignorance might infect other regions, too.[17] For the Convention, the peasants of the west were feeble, superstitious, 'a backward people'; as for their leaders, they were an ill-assorted mix of 'fanatics, refractory priests and émigrés' who were among the most 'committed enemies of the republic'.[18]

The term 'brigandage' was widely used to describe the tactics of the guerrillas. Differentiating between soldiers and civilians was judged essential to the orderly conduct of war; it was what identified an army and gave it the right to bear arms. Soldiers were subject to military discipline; civilians were not. And the fact of wearing a uniform gave men rights and obligations that they could never aspire to in civil society, privileges as well as duties: the right to bear arms in wartime, to be judged by their peers, to defend themselves with steel, and to kill their enemies. If they were deprived of their status as soldiers, they risked being treated as common criminals, accused of acts of violence, assault, murder, and pillage; indeed, in peacetime soldiers did face such charges, as when they attacked civilians, pillaged farms, or raped local women. Military justice maintained a distinction between acts of war and war crimes, as is shown in the many hangings of its own men that the army ordered during the Thirty Years War in Germany. And when they escaped from military regulation, soldiers risked crossing the line of tolerance and being classed as outlaws or brigands, handed over to local magistrates, and exposed to the full rigours of the law. This happened frequently when men deserted from the ranks and sought to find their way home through places they barely knew and where they felt few bonds of familiarity with local people. Tired, starving, and desperate, they were often reduced to petty crime to stay alive.[19]

The Republic's insistence that the Vendean was not a real soldier, but little more than a brigand, was not without consequences. It deliberately set out to criminalise him and to strip him of the sense of military honour that is one of the motors of all military societies and which, since the Middle Ages, had been an essential contributor to the soldier's self-esteem. In that sense, it sought to diminish him, to question his valour and his masculinity at a time when military honour and masculinity were doubly significant, since they were also essential images in the representation of republican virtue.[20] It also left him with no recourse to the law or the public authorities, and therefore prey to any attack or reprisal, naked and alone in a savage underworld on the margins of society. By using the imagery of brigandage, the republicans quite knowingly attacked the integrity of those ranged against them; like Joseph Bonaparte in Spain a decade later, they painted them as enemies not just of the political regime, but of the entire social order. Charles Esdaile cites the *Gazeta de Sevilla*, a paper known for its loyalist views, in January 1812. The partisans, it insisted, were given to attack everyone, indiscriminately. 'Spaniards, Frenchmen, Englishmen, all are the victims of their perversity: priests, mayors, day labourers, muleteers, no one is spared ... such are the instruments that are now at the service of perfidious Albion.'[21] They were not soldiers, the republicans insisted, but men lurking in the shadows, seeking refuge in village society; and by making it impossible for the French to distinguish between soldiers and civilians, they prevented the troops from carrying out the laws of war. Yet they were wholly dependent on the guerrillas for their defence, as the regular army had ceased to defend them and had become the army of their oppressors.[22]

For the revolutionaries, the guerrillas were men without virtue, without morality. If we examine republican rhetoric, there was a huge difference between *la grande guerre* and *la petite*; guerrilla tactics were a rejection of any suggestion of chivalry; and guerrilla warfare was fundamentally unjust, unable to appeal to either *ius in bello* or *ius ad bellum*. To delegitimise them, and also to justify the harshness of the repression that followed, the fact that no distinction could be drawn between soldiers and civilians was used to justify the scorched earth policy that the Convention demanded. The consequence, as we know, was the devastation of large tracts of land in 1794, in a region officially described as being 'infested with brigands' who enjoyed the protection of local people.[23]

The Republic had no reason to show them any indulgence and sent a punitive expedition against them:

> They burned, lamented one observer, all the hamlets and cottages, massacring some of those labouring in the fields, burning corn and straw in the barns and in the farmyards, slaughtering and devouring a large number of sheep and cattle, and carrying off or slaughtering all the horses and mules.[24]

Bertrand Barère tried to justify this approach by referring his listeners to the precedent set by the devastation of the Palatinate on the orders of Louvois in 1689; the Vendée, he declared, was 'the Palatinate of the Republic'.[25] But for the victims the devastation of their fields and the massacre of so many villagers could only be understood as an atrocity, if not a war crime or – to use the most extreme vocabulary – a case of genocide.[26]

But what was the basis of their charge that the Vendean war lacked legitimacy? Was it the way in which the rebels conducted themselves, the nature of the warfare they unleashed? Or was it the fact that they were rebelling against a legally constituted government? During a civil war it is easy to confuse these two elements. And in using such criminalising language the revolutionaries appeared to authorise this confusion, attacking at one and the same time their method of fighting and the unenlightened values which – in revolutionary eyes – they encapsulated. This proved important, since in other contexts guerrilla tactics had been justified by the French and had even been recommended to French troops themselves. During the Dutch wars, for instance, Louis XIV had sent into his frontier fortresses *compagnies libres* composed of partisans who specialised in guerrilla actions, while regular troops were periodically detached from the normal duties to attack convoys and enemy escorts.[27] And in 1814, during the *campagne de France*, Napoleon himself would appeal to the people to rise against the enemy in a partisan movement for self-defence.[28] In France, as in all other European countries, *petite guerre* was tolerated when circumstances demanded it, and since the time of Louis XIV there had been a number of attempts to integrate guerrilla actions into French military strategy.

Small wars were recognised to be useful under certain circumstances, and, like marauding and the *dragonnades* of the seventeenth century, their legitimacy was accepted by military strategists in wartime. Indeed, a number of influential works had been published on the subject. These included eighteenth-century publications that gave regular army officers advice about what one of them, *Le Partisan, ou l'art de faire la petite guerre avec succès selon le génie de nos jours*, referred to, quite unashamedly, as 'the part of the Military Art which I discuss'. And what was the purpose of recourse to *petite guerre*? It was best, claimed the author, to assign to guerrilla activity a unit of troops that was linked to the regular army without being fully integrated into it. 'This is a unit of light, mobile troops, between a hundred and two thousand strong, and to be kept apart from the Army; it should be used to secure the route and the camp; to reconnoitre the country and check on the enemy positions; to take out their posts, their convoys and their escorts; to mount ambushes against them, and use all the tricks that can take them by surprise or spread fear among them.'[29] In this passage the author goes little beyond the analysis made by the *Encyclopédie* in the 1760s, which saw guerrilla operations as having a very specific role to play in modern warfare. *Petite guerre*, it declared, 'is war made by small detachments or marauding parties, whose object is to uncover the tactics of the enemy, to observe the movements of their army, to get in their way and harass all their operations, to surprise their convoys, to exact contributions, and much more'. The author continued, 'This kind of war calls for a great deal of intelligence and ability from the commanding officers. They must be able to distinguish the strengths and weaknesses of the enemy army's position, and judge the advantages which the nature of the terrain can offer for attack and surprise, either while it is on the march or out foraging. They must also know how to subvert the enemy's plans by its own movements, and to take precise observations so as not to be deceived by false movements.'[30] The *Encyclopédie*, like the military manuals that were devoted to the question of *petite guerre*, speaks of it as a legitimate tactic, and one which the French army has every interest in mastering and exploiting. Collectively, these texts serve to remind us that there was a wide gulf in the second half of the eighteenth century between what was perceived to be legal and illegal in matters of military strategy. They remind us, too, that guerrilla warfare was not always denigrated by military strategists as it came to be in the Vendée.

Another military writer of the period, Grandmaison, wrote at length in the benefits that could be obtained from using guerrilla tactics. There could be various objectives in sending irregular troops into action. 'Sometimes it is to learn news of the enemy, and to worry them in as far as conditions permit; at others we might want to surprise a guard post or a detachment of men, or attack a convoy or the enemy rearguard, or to fall on their equipment, or on their foragers as they file past with their horses to go marauding; or else there are times when we want to impose taxes or execute more long-range objectives.'[31] In all these circumstances, it would seem, the military authorities saw guerrilla fighting and skirmishing, when attached to the work of regular battalions, as perfectly legitimate tactics, whether they were directed against the enemy army, used for policing or peacekeeping, or to sow terror and panic in the civilian population. Another writer, Jeney, makes no effort to conceal this terrorising dimension, which was widely used in Old Regime armies. Indeed, he implies that to subjugate a civilian population it is often necessary to start by spreading terror. So when the army stops near a village or hamlet, the first thing to do is to drag the principal citizen of the community before the military authorities. 'He will be threatened with death and his house with being torched, should any of the villagers try to escape before being released; this scene should always take place in public, in the presence of all the villagers.'[32] He describes this without any apparent embarrassment; it was part of a recognised procedure that was seen as necessary if the army was to extract any information from a reticent population.

Yet at the same time he listed measures that should be denounced and punished if they were used by the enemy in their manoeuvres against the French. This is never an even-handed exercise. What on the one side would be seen as brigandage or even terrorism is justified by the needs of the army and the necessities of war.

'Brigandage' was a complex and multilayered construct, an amalgamation of the anti-revolutionary and the antisocial, as well as a refutation of the tactics used by the other side. It was not, of course, used only in the west. The Revolution was faced with guerrilla warfare along many of France's borders, especially where it was in conflict with rural communities, or where the opposition sought to take advantage of its knowledge of the landscape and the terrain. Some of those who were denounced as brigands, like the *barbets* along the Italian border, seem to have set out on their paramilitary career with largely idealistic goals, inspired by a political or ideological programme, before falling into lives of crime in order to feed themselves and keep their band alive. But after a generation at war there was often little to distinguish them from the most banal of thieves, and the revolutionaries had no interest in drawing fine distinctions between men who; in their eyes, seemed unworthy and dishonourable. All were presented as bandits, brigands, outlaws; all shared a collective guilt. In the Vendée their initial presumption was that local peasant fighters had been misled by men who were committed to brigandage. The deputies Villers and Fouché addressed the rural population and sought to separate them with their supposed naivety, from the more evil, corrupting 'brigands' who had 'seduced' them. Two years later, with reference to the 'prodigious number of women and children who follow the Vendean army', their successor, Laplanche, was ready to denounce the entire population of the Vendée as 'a race of brigands'.[33] The word 'brigand' had become a general term of rejection, used to describe anyone who was thought dangerous or unruly or who, like army deserters, were armed with knives or muskets, often the service weapons they had stolen when they fled from their regiment.

For the revolutionaries, then, the Vendeans had become common criminals, no more than the most dangerous of the many brigands they encountered during these years, men denuded of all honour or decency, men to be punished as traitors and if necessary shot down. The guerrilla fighter had not entered into any contract with the state or with civil society. He was therefore a private citizen, acting in his own interests, often for his own material advantage. Unlike the soldier who had accepted to sacrifice himself for the common good, the guerrilla was a bandit who had cast himself outside civil society.[34] His protestations that he was fighting in a cause he believed in – for the king or the Catholic Church – fell on deaf ears. This was a murderous civil war – as the figure of 140,000 to 190,000 dead among the inhabitants of the region bears witness – and for the Republic the insurrection was seen as a simple act of treason, committed in the midst of a major European war by religious fanatics and royalist extremists led into error by the local nobility and by an unrepresentative and intransigent clergy.

The situation was made worse by local hatreds and rivalries. The Jacobins of the western cities had hurled insults and republican invective against the rebels from the very first hints of a revolt, regarding them as murderers who had chosen the path of treason and had placed themselves outside the law. For them, 'brigand' became a natural term of abuse, and the rebels in their turn adopted it with pride, placing it on their escutcheon as part of their collective identity. They were, like the Patriots of Avignon in 1790 or the Belgian peasants who revolted against the French authorities in 1798, 'bons brigands', leading armies and representing their communities and their values against an outside invader. The term was adopted with pride and a degree of swagger – illustrative of a certain bravura, suggestive of an alternative patriotism, that of their region, their culture, their people, in defiance of the revolutionary state. With, of course, a marked and publicly sported antipathy – which was heartily reciprocated – for the towns of the west, with their bourgeois, their public officials, their republican dogma. In the rural west there is good reason to believe that the citizens of towns like Saumur and Angers, Poitiers, and Nantes, were detested far more heartily than any distant leader in Paris. Against them and their interventions in the countryside – whether to collect taxes or requisition troops or supplies or horses for the army – the irregulars of the Army of the West would fight with an especial relish.

The revolutionaries used a whole battery of rhetoric against them. Before the tribunals, which were set up across the west, no distinction was allowed: all, leaders and followers, chiefs and partisans, officers and men, were condemned as brigands, as outlaws, guilty of the capital crime of rebellion against the Revolution and against the French people. In the process, the entire population of the west became demonised, represented as a seditious Other; as fanatical, obsessed with nobility and feudality, manipulated by their *seigneurs*, and paid by the agents of perfidious Albion. On the battlefield, moreover, they were accused of levels of cruelty that came close to barbarism, a cruelty that affected the whole spirit of the war and condemned both sides to depths of inhumanity. Turreau maintains that in the Vendée the rebels put to death in atrocious circumstances any republican soldiers whom they had waylaid or ambushed. But he also admits that the republicans contributed to the atmosphere of mutual hate through their own policy of giving no quarter. The only difference, in his view, was that the army was content to shoot the rebels they captured, whereas the Vendeans tortured their victims first.[35]

In civil war, hatred is the dominant emotion, the desire to avenge oneself and to rid the country of a virus, an infection. So it was insufficient for their opponents to depict the Vendean as a simple brigand: the brigand, after all, may be a criminal and may deserve the death penalty, but he is still a human being, capable of thought and emotion and able to make rational decisions; even if, as a Vendean, a backward and superstitious example of the race. But could men who committed such outrages and atrocities really be deemed to retain the slightest vestiges of humanity? In the revolutionary mentality, he had not by his actions and his refusal to recognise 'the rules of a regenerated humanity; that is to say of a humanity that had rediscovered its true nature', forfeited his right to be considered a human being? For Sophie Wahnich, 'the logic of exclusion brought together a theoretical humanism (for it is in the name of humanity that he agrees to act) and an anti-humanism in fact (since the life of a man, as of a people, is worthless if he betrays his humanity), which is translated in the actual condition of France into a radical civil war'.[36] In this way, the tactics of the guerrilla combined with ideology and the bitterness of civil war to present a dialectic vision of a struggle between archaism and modernity and to strip the enemy of the last vestiges of his humanity.[37]

In the words used by the commissioners and deputies on mission sent into the west, we can see how the expressions they reserved for the rebels sought to dehumanise them, to compare them with wild beasts and identify them as wolves rather than men. We find the same words used over and over again in republican reports on the west, where the Vendeans are routinely described as *'bêtes'*, *'monstres'*, *'vermine'*, *'animaux féroces qui cherchent à dévorer la république'* – words that are evinced with a cold deliberation in order to devalue the lives of their adversaries, to relegate them to some subhuman species. They had, it was implied, a taste for blood and for butchery that belonged more naturally to the wild beasts of the forest and, like the refractory priests whose ideas they so willingly swallowed, they reserved their bestiality for their republican victims alone. What is more, their habits and movements had come to resemble those of the wild animals among whom they lived. They were depicted as men in the guise of beasts, hiding behind hedges before 'leaping' from the bushes on their unfortunate victims or retiring to their 'lairs' to recover from their battles and 'lick their wounds'. This discourse had, of course, huge propaganda value to the Republic, especially in the republican strongholds of the interior and in the army, where the soldiers were apprised of the cowardly and vicious character of those against whom they were fighting – men without honour whose few successes could always be explained as the result of brute force and animal desperation. As a cavalryman from Coutras in the Gironde wrote in a letter to his parents, the civil war almost necessarily had a dehumanising effect on the peasantry, since 'if they had not acted like wild animals they would surely have been wiped out'.[38]

Both the government and the army knew the value of winning over public opinion to their cause, and they exploited these images with persistence and ingenuity. Once he had been divested of his humanity and depicted as a wild beast, it was easier for the Vendean to be made a target for popular anger and reprisals. Civilians would be less prone to offer him shelter, and revolutionary tribunals would have fewer scruples about sending him before a firing squad. Wild beasts evoke fear: few in rural France would have had any qualms about shooting a wolf or a wild dog, or any animal that posed a threat to their homes or their flocks. Indeed, many peasants would have had the experience of taking part in a hunting-party to rid their village of a roaming wolf. And wolves could easily be confused with monsters; it was, after all, less than 30 years since the Bête du Gévaudan had spread panic in the pasturelands of the Lozère, where he had, it was said, picked out the most succulent local children for his Sunday lunch.[39] What this animalising discourse achieved was to deny the Vendean his status as a human being, and to provide the ultimate justification for terror and destruction. It was a language of extermination. The republicans were not wanton killers: they killed so that they should not themselves be killed. And the killing became an exercise in cleansing, an act of purification. All ideas of compromise became unthinkable, creating a brutalised mentality and making it very difficult to restore peace to the war-scarred parishes of the west. In 1795, and again in 1800, these conflicts would be easy to rekindle.

Just as they would do later in Spain, the soldiers sent into the Vendée learned from republican propaganda that their adversaries were inferior creatures, a subhuman race to whom it was pointless to offer the honours of war. They did not hide their contempt for Vendean tactics, or their anger, when faced with enemy atrocities. Nor did they conceal their sense of inherent superiority in the face of what they interpreted as primitive superstition. For the *commissaire aux armées*, Jean Savary, religion lay at the root of all the excesses he encountered in the west. The Vendean believed any number of absurdities that were whispered to him in the confessional box, in the process becoming 'the instrument and the victim of those who made him do what he did'.[40] For Savary, this was the only reasonable explanation he could suggest for the cruel and sadistic nature of the men he encountered, and it goes far to explain the viciousness of his own response. In the words of a municipal officer from Bordeaux serving with the Army of the West in 1793, 'in the Vendée the idea that one might pardon an enemy is unthinkable'. And he went on to insist that '40,000 brigands' were steeped in republican blood; 'some of these cowardly slaves, believing that they might save their own skins by surrendering were made to realise that they had made a great mistake, for no sooner than they had fallen into our hands than we despatched them to join the moles below.'[41]

In this context he had no hesitation about shooting prisoners – after the fall of Noirmoutier they would be shot in batches of 1500 at a time – and priests always figured among their most hated enemies. For Jean-Charles Bouquet, the son of an ironworker from Reims and a devoted republican, the west was a country infested with priests, and he did not hide his pleasure at seeing them drown in the Loire:

> The priests who dared to resist the general will were put on board a ship with those refractory priests who were already captured, but, through the intervention of the Genius of the Republic a rotten plank became detached, the vessel shipped water on all sides, and sank with its entire cargo of evil priests.[42]

His words may seem suggestive of an almost inhuman cruelty. But it should never be forgotten that these soldiers were the same men who served in Belgium or along the Moselle, and that in these less torrid contexts they had never thought of vengeance of this kind. They were not naturally savage or bereft of humanity, but their sensibilities had been hardened through their service in the west, by incessant republican propaganda, but also by direct experience of this 'small war', with its banal acts of cruelty and propensity to torture. They saw themselves as avengers, whose anger had been roused by the cruelty of civilians – women among them – towards their comrades, and by the acts of torture to which republican troops had been subjected. André Amblard spoke of his own reaction to news of the massacre committed by the Vendeans at Machecoul. According to him, they had massacred pitilessly anyone who had taken the side of the Republic, and he continued:

> Tired of butchery, or as a refinement to their cruelty, they took 500–600 of the prisoners, tied them together and led them to a piece of high ground where there was a windmill, and there, on a grassy open space, they battered them with iron bars. Everyone joined in the beating, women and children among them. The poor wretches who were attacked uttered the most searing cries.[43]

Recognising that his account might seem partisan and that his readers might need some convincing, he added various corroborative details, assuring them that it was an 'incontestable truth' that had been witnessed by one of his officers who had seen the dead bodies and who had 'married the widow of a lawyer who had been among the victims'. The officer, Captain Fayolle, and his wife had given Amblard details of the massacres that were 'even more terrible than those I've described'.[44] Amblard and his comrades were in no doubt as to the truth of their account: for them Machecoul was a cowardly massacre, the atrocity which, more than any, had unleashed the bloodbath in the west.

For the army their opponents were subhuman, beings who did not deserve to belong to the human race, and the propaganda against them constituted an essential part of the war effort. But it also contributed to make the war unremitting and often barbaric, a war where regular troops, every bit as much as the irregulars they faced, lost many of their sensibilities when faced with death and killing. Some would talk of their own involvement in such butchery more openly than others, and some also confessed to doubts. From the village of Montglonne, near Saint-Florent-le vieil, a soldier admitted in a letter to his mother that he was repelled by what he had to do. Every day, he admitted, they would shoot batches of prisoners. 'They bring them to us on a daily basis', he wrote, adding that they had 'put a bullet in each of them'.[45] He recognised that he had acted as a killer, yet his letter did not express any evident regret, or moral discomfiture. They were 'brigands', after all, fighting a *petite guerre* in which they abused and mistreated the French army, and this in a civil war in which they challenged the Republic, the state, the very people of France. Their role was more that of criminals than soldiers, and for that they deserved no sympathy. It is further evidence that this was a war fought with rhetoric as well as arms, and that the rebels' resort to guerrilla tactics was a potent weapon to be used against them. The republican victory was that of one discourse over another, and that of an urban, laicised culture over rural autonomy and religious conservatism. The work of propaganda, in this civil war between two very different ideas of France, proved as important in securing victory as the work of the battalions on the battlefield.

Notes

1. Muraise, 'L'insurrection royaliste de l'Ouest', 72–3.
2. Quoted in Girard, *Pourquoi la Vendée?*, 11.
3. Martin, *La Vendée de la mémoire*, 34.
4. Tilly, *The Vendée*; Bois, *Les paysans de l'Ouest*; Tilly, 'Analysis of a Counter-Revolution'; Petitfrère, 'Origins of the Civil War in the Vendée'; Mitchell, 'Resistance to the Revolution in Western France'.
5. Petitfrère, *Les Vendéens d'Anjou*; *Les bleus d'Anjou*.
6. Balzac, *Les chouans*.
7. Martin, *La Vendée de la mémoire:* v.
8. Martin, 'Vendée, guerre de', 502.
9. Picaud-Monnerat, *La petite guerre au XVIIIe siècle*, 42.
10. Turreau, *Mémoires*, 33.
11. Ibid., 199.
12. Peyrard, *Les Jacobins de l'Ouest*, 285.
13. Martin, 'Introduction', 10.
14. See Martin, *La Vendée de la mémoire*.
15. Tilly, 'Local conflicts in the Vendée'.
16. Aulard, *Recueil*, 5: 98. For a more detailed discussion of republican discourse on the Vendée, see Tyson, 'The role of Republican and *patriote* discourse', 79.
17. Aulard, *Recueil*, 4: 445.
18. Dupuy, 'Ignorance, fanatisme, et contre-révolution', 38–9.
19. Forrest, *Déserteurs et insoumis*, 138–68.
20. Clark, 'The Rhetoric of Masculine Citizenship', 6.
21. Esdaile, *Fighting Napoleon*, 18.
22. Walzer, *Just and Unjust Wars*, 179–80.
23. Rowe, 'Civilians and Warfare during the French Revolutionary Wars', 113.
24. Secher, *Le génocide franco-français*, 206.
25. Weber, 'La stratégie de la terre brûlée', 193.
26. See especially Pierre Chaunu, who, in the introduction to the book by Reynald Secher, cited above, writes: 'Cette guerre fut la plus atroce des guerres de religions et le premier génocide idéologique'. Secher, *Le Génocide franco-français*, 24.
27. Lynn, *Giant of the Grand Siècle*, 542.
28. Lentz, *L'effondrement du système napoléonien*, 550–2.
29. Jeney, *Le Partisan*, 1–2.
30. 'Petite guerre', entry in *L'Encyclopédie*, 12: 466.
31. Grandmaison, *La petite guerre*, 111.
32. Jeney, *Le Partisan*, 46.
33. Viguerie, 'La Vendée et les Lumières'.
34. Forrest, Alan. 'The Ubiquitous Brigand'.
35. Turreau, *Mémoires*, 31.
36. Wahnich, 'La logique de l'exclusion révolutionnaire', 74.
37. McMahon, *Enemies of the Enlightenment*, 202.
38. Pagès, 'Lettres de requis', 155.
39. Smith, *Monsters of the Gévaudan*, 1–26.
40. Savary, *Guerre des Vendéens*, 1: 29.

41. Forrest, Alan. 'La guerre de l'Ouest', 98.
42. Coutansais, F. 'La guerre des Géants vue par les Bleus', 4.
43. Amblard, 'Les guerres de Vendée', 66.
44. Ibid.
45. Letter from Feugier to his mother in Valence (Drôme); the letter is part of the private collection of M. Guy Bahuault, who was kind enough to allow me read and to use it.

Guerrillas and bandits in the Serranía de Ronda, 1810–1812

Charles Esdaile

Department of History, University of Liverpool, UK

The Spanish Guerrilla (1808–1812) which has given its name to ideologically motivated insurgencies is usually portrayed as a patriotic uprising against the French occupation forces of Napoleon. It was that, in part, but also many other things besides. This case study illustrates its overlap and convergence with banditry but also with social unrest turned into uprisings directed by poor Spaniards against their creditors, as in the storming of Ronda by insurgents in 1810. From the propaganda of the day to the subsequent Spanish patriotic historiography, there has been a tendency to exaggerate the amplitude of events and also the damage that was done to the French forces and the casualty figures inflicted on them.

If any country ever needed a 'good-news story', it was Spain in the summer of 1810. In November 1809 the chief Spanish armies had been shattered in the battles of Ocaña and Alba de Tormes; in December 1809 the stronghold of Gerona had finally succumbed after an epic siege that had lasted six months; and in late January 1810 the French had overrun Andalucía in a three-week blitzkrieg. To military humiliation, meanwhile, had been added political disaster in that the Junta Suprema Central, the provisional government that had been formed following the uprising of 1808, had been overthrown by a coup. Since then, true, the situation had improved a little: Cádiz – Spain's third capital in three years – had proved invulnerable to French assault; Badajoz had stood firm in the face of an attempt to bully it into surrender; and governmental authority had been restored through the creation of a council of regency and the convocation of elections to a new national assembly. Yet the dark clouds that beset the country had scarcely been dispelled: if, indeed, they could do even so much, what remained of the Spanish armies could clearly hope to do no more than hang on to such territory as remained to the Patriot cause; the Anglo-Portuguese army of Lord Wellington had in effect withdrawn from the conflict in Spain; and the first news was starting to arrive of a series of revolutions in the Spanish colonies, with the loss of Andalucía, now the financial mainstay of the war. To repeat, then, what was needed was a 'good-news story', and that in the shortest possible of orders: as

Spain was waging a people's war, it followed that morale had to be kept up at all costs.

All this being the case, when news arrived that a major insurrection had broken out in the Andalusian mountain range known as the Serranía de Ronda, it was seized on as an absolute godsend. The official gazette was quick to notice the story while it was soon being featured in many of the newspapers that had quickly become a central part of Cádiz life. Here, for example, is an excerpt from a long report that appeared on 24 August 1810 in *El Observador*:

8 August. On this day an enemy force that was occupying the town of Jimena attempted to send out a detachment of 200 men to seize the harvest from the cultivated area known as the Huerta de los Granados, but, opening fire on the enemy column from the other bank of the river, the Patriots of Gaucín managed to force it to retreat. At the ford at Colmenar, meanwhile, a party of eight Frenchmen crossed over to pillage the estate known as Las Abiertas, but a group of Patriots resisted them and forced them to retire. In this last action, whereas none of our side were lost, three enemy soldiers were killed, one of them a Polish cavalryman, the mare that the latter was riding being secured by the man who killed him, Bartolomé Román Carrero ... Finally, the public will learn with satisfaction of the actions of an elderly resident of Gaucín named Francisco Lermos Fernández. Thus, moved by nothing more than ardour to defend the just cause, Lermos set aside the fact that he was a chronic invalid and sallied forth from his home armed only with a fowling piece with a broken lock. Accosting a Polish cavalryman, he aimed his gun at him with the cry, 'Surrender or die', whereupon, seeing that he was fairly caught, the soldier threw down his sabre and dismounted from his horse, Lermos then taking charge of both, which he has since retained as his own property ... and conducting his prisoner to the town of Casares.[1]

The interest that was shown in this story, meanwhile, is entirely understandable. In the first place, situated no more than 50 miles away from Cádiz, the Serranía de Ronda enjoyed easy communications with Cádiz via the port of Tarifa, which, like the capital, had remained secure from the French onslaught, and therefore offered a source of news that was readily accessible. In the second place, it really was a 'good-news story' in that the victorious French finally seemed to have met their match. In the third place, the war in the Serranía served a teleological function in that it showed that Spain was not yet lost, or, to put it another way, that even seemingly defenceless civilians armed, at best, with improvised weapons could still strike a blow for freedom, and not just strike a blow for freedom but vanquish the enemy, thereby, incidentally, gaining access to a variety of valuable booty. In the fourth place, there were many elements in Cádiz that were eager to emphasise the role that was being played by the people in the Spanish war effort as a means of legitimising the radical programme of political reform that was ultimately to form the basis of the constitution of 1812. And, finally, in the fifth place, the place of the war in the *serranía* in the news was perpetuated for some considerable time by the bitter disputes that erupted among its leaders and were acted out in a series of pamphlets and manifestos that appeared in Cádiz over the course of the next few years.[2]

If it is an exaggeration to say that the deeds of the *serranos* had Cádiz buzzing with excitement, the reports of British visitors to the city certainly suggest that the coverage which they received created a very strong impression.[3] A legend in their own lifetime, then, the *serranos* were not forgotten after 1814, not least because several French soldiers who had fought them left dramatic accounts of their activities.[4] Thus encouraged, many of the great nineteenth-century chroniclers of the Peninsular War were full of praise for the exploits of the *serranos* while the tradition has persisted down to the present day.[5] So much for the historiography, then. In this article we shall obviously re-examine the traditional version of events and test it out against a number of sources that rarely figure in the writings of those who have sought to celebrate the resistance of the *serranos*. Before going any further, however, it is probably worth saying a little more about the district of which we are speaking. In brief, then, the Serranía de Ronda is a roughly oval-shaped mountain range stretching north-eastwards from Gibraltar to Málaga. At its heart lies the city of Ronda, but in 1810 this had but 15,600 inhabitants, the rest of the populace living in 20 or more *pueblos* of which the most important were Casares, Jimena de la Frontera, Cortes de la Frontera, and Gaucín. Intersected with narrow valleys, the mountains rise to a height of perhaps 5000 feet, while they are honeycombed with caves and cloaked with forests of pine, ilex, and cork. With the only roads no more than winding goat tracks, the district is clearly perfect territory for a guerrilla war.[6]

Needless to say, this fact was as clear in 1810 as it is today. Many of the smaller *pueblos* in the region having remained free following the occupation of Ronda itself on 10 February, hardly had the French reached the Mediterranean coast than a variety of figures in the Allied camp therefore had come to realise that the rugged mountains of the district were an ideal base for guerrilla warfare. At all events, as early as 6 February 1810, the commandant general of the Campo del Gibraltar, Adrián Jacome, commissioned a supernumerary officer named Francisco González Peinado to range the interior of the Serranía with the aim of raising the inhabitants in revolt. Accompanied by a small group of regular army officers and the parish priest of Algeciras, Peinado duly visited such *pueblos* as Algatocín and Jubrique, while he was very quickly joined by a small British mission that had been dispatched to the area for the very same purpose by the governor of Gibraltar, Colin Campbell.[7]

These emissaries found the area in a state of considerable ferment, for, whereas the invaders had behaved reasonably well in the larger towns and cities of Andalucía, in more remote districts such as the Serranía the inhabitants had in consequence been subjected to considerable brutality as the French penetrated the region in the wake of the fall of Seville. With the area possessing a long tradition of banditry and smuggling, however, violence was met with violence, and the result was that González Peinado and his fellow agents were able to secure a ready hearing, while they were also joined by various local notables, of whom the most important was José Serrano Valdenebro, a 68-year-old retired naval officer who had been living in Gaucín.[8]

On the surface, popular protest had therefore been smoothly assimilated by the legitimate authorities, and this in turn meant that it could be channelled into structures that ensured that it did not challenge the established order. At the behest of González Peinado and Serrano Valdenebro, then, large numbers of *serranos* were soon massing for an attack on Ronda, while, quickly designated by the regency as the head of the insurgents, the latter organised the populace into a home guard based on the concept of the parish company and officered by a variety of local notables, most of them either priests or friars or representatives of the propertied classes.[9]

To continue with the traditional account, what all this meant was that in the Serranía de Ronda 'people's war' did not conform to the same pattern as that seen elsewhere in Patriot Spain. Thus, the inhabitants did not choose the classic guerrilla band as their preferred mode of resistance to the invader, but rather allowed themselves in effect to become an extension of the armed forces of the state that was mobilised at the behest of the authorities in connection with particular military operations, whether these aimed at attacking enemy garrisons such as those established at Ronda or Gaucín, impeding the movement of French columns, or defending *pueblos* threatened by pillage or reprisal.

In retrospect, of course, this can be seen as a very useful construct: with Andalucía a region that was marked by some of the sharpest social divides and the worst poverty in the whole of Spain, the idea that the populace might take up arms on its own account was an idea that was particularly frightening for the local élites. That said, it was not just a construct: on the contrary, the historical record shows that the form of warfare that has been outlined was by no means uncommon in the *serranía*. In so far as an attack on a French garrison is concerned, even if Ronda was in the event abandoned by the enemy without any resistance (see below), we need only cite the attack on the town that we have just mentioned. In so far as the harassment of French columns is concerned, we have a variety of lengthy first-hand accounts written by Serrano Valdenebro, the French hussar officer, Albert de Rocca, and, finally, the senior British commissary, Richard Henegan.[10] And, finally, in so far as the defence of a *pueblo* is concerned, we can point to the furious battle that took place at Algodonales on 2 May 1810 when it was attacked by a French column commanded by General Maransin in an action that was much trumpeted by the *josefino* authorities as the epitome of what the populace could expect if they were seduced into taking up arms against the invaders.[11]

So far, so good: the *serranos* appear as a people in arms fighting heroically for the cause of *dios, rey y patria*, the image being strengthened still further by the frequent mentions that can be found of women taking part in the fighting in the exact same style as some of the images in Goya's famous *Desastres de la Guerra*. Indeed, so strongly are many of the images of combat portrayed in the series redolent of the idea of warfare *a la serrano* that one cannot but wonder whether it was press reports of the fighting in the *serranía* that lay at the bottom of the manner in which the artist visualised the struggle.

Yet in reality matters are by no means so simple. In the first place, it is quite clear that the legitimate authorities – González Peinado and Serrano Valdenebro – did not enjoy a monopoly as far as resistance to the enemy was concerned. On the contrary, a variety of guerrilla bands sprang forth directly from the ranks of the populace, one such being the one formed by a shepherd from Ubrique named Pedro Zaldívar that became known as *la partida de Palmetín* (literally, 'the Palmetín gang'), and another group formed by the rather unlikely person of Andrés Ortiz de Zarate, a self-styled professor of mathematics who had taken refuge in Gibraltar and joined the mission sent out to the Serranía de Ronda by General Campbell as an interpreter.[12] The commissions in which such an action resulted serving as a useful reinforcement of their position, not least because it gave them access to the shipments of arms and ammunition regularly dispatched to the *serranía* via such places as Algeciras and Tarifa, the chieftains that headed these bands frequently applied for the sanction of the civil or military authorities. Even when they elected to act in this fashion, however, it is clear that they remained freelance operatives who engaged in combat with the invaders (if, indeed, they did so at all) on their own account and remained apart from the civilian population as a whole. For a portrait of what may or may not have been a typical band of irregulars, we may turn to a band of 48 men that Richard Henegan encountered in March 1811:

> The genuine guerrillas differed materially in appearance from the *serranos*: the latter were not under the command of one chief, as were the guerrillas, but acted each on his own responsibility. At one moment cultivating his land, at the next flying to the signal that called him to the mountain strife. Ever ready on the hills to wreak his vengeance on the French, ever ready on the plains to watch over his paternal property. Their dress also was different. The guerrilla in his short jacket of russet brown, and leather leggings in the same dark colour, was wanting in the jaunty smartness that characterised the *serrano*. The latter was almost invariably clad in velveteen of olive green, profusely ornamented with silver buttons and laced with ribbons of many hues. A tight white stocking enclosed a well shaped leg, over which *botines* of curiously wrought leather completed its symmetrical appearance, and a small *sombrero*, shaded by its plume and saucily stuck on one side, gave the last finishing stroke to the *serrano*'s appearance. The guerrillas always wore a broad leather belt, amply furnished with murderous weapons which were generally the trophies of success, their own arms being discarded and replaced by the more highly finished ones of the French officers they killed.[13]

More mobile, embodied on a more-or-less permanent basis, and, in all probability, better armed than Serrano Valdenebro's home guards, it was beyond doubt the guerrilla bands that bore the brunt of the struggle against the French. Such, at least, is the impression that is derived from the French sources. Here, for example, is the hussar officer, Rocca:

> Bodies of Spanish partisans, three or four hundred strong, scoured the country on all sides ... These troops of partisans or guerrillas served to keep alive the fermentation of the country, and kept up the communications between Cádiz and the interior of Spain. The people were led to believe that the Marquis de la Romana had

beaten the French at Trujillo, and that the English in a sortie from Gibraltar had completely defeated them near the sea. These reports, however improbable, being skilfully spread, were received with transport.[14]

And here the commander of the French forces in Andalucía, Jean de Dieu Soult:

Driven by misery and the desire to pillage, almost the whole of the insurgents and smugglers of the mountains between Ronda and the Campo de San Roque ... have come down into the plains and committed a number of excesses. Several mobile columns have been sent to hunt them down, but they have only been able to drive them away momentarily, and have not taken many prisoners.[15]

In short, classic guerrilla warfare – the guerrilla warfare of the sort associated with bands of irregulars – was part of the war in the *serranía* after all, and this in turn means that it is difficult not to question the traditional version of events. More damning, still, however, is a source that has only recently begun to be interrogated in the form of the extremely detailed tables of France's losses in the Napoleonic Wars that the French military historian, Aristide Martinien, published in Paris in 1908.[16] For reasons that cannot be fathomed, these tables, which are generally agreed to be unrivalled in their accuracy, have never been applied to the guerrilla struggle in Andalucía or, indeed, anywhere else, and yet doing so throws up a series of statistics that seem to bear no relation to the picture we have thus far drawn in this article. According to the traditional account, the guerrilla war was a bloodbath in which the French lost hundreds of men a day due to guerrilla action. Martinien, however, suggests a very different picture. Thus, between February and August 1810 – the period covered by the accounts of combat mentioned earlier in this article – the number of French officer casualties in the area of Ronda amounted to just 10 dead and 23 wounded. We may therefore assume that, in something over six months of what is supposed to have been the most bitter fighting cost the French no more than, at the most, 500 casualties. What is worse, no casualties at all are recorded for either the actions described by Rocca and Serrano Valdenebro or the sack of Algdonales.[17]

With respect to the first model of resistance that we have discussed, it is not just the issue of casualties that should give us pause. On the contrary, from quite early on in the story of insurrection in the *serranía* we come across hints that all was not as it should be. At first, the disturbances that broke out in the area after it was occupied in February 1810 aroused much hope among the Patriot authorities. 'The uprising of the Serranía de Ronda is a glorious affair,' wrote the head of the council of regency and victor of Bailén, General Francisco Javier Castaños. 'Put on the right footing, it could be of major help to future operations.'[18] However, sent to get the *serranos* together in order to use them to assist the operations of a division that was to be disembarked from the sea under Luis Lacy, Ambrosio de la Cuadra found that 'the right footing' was not something that was to be attained with any ease. In the first place an interview with the supposed leader of the rebels proved a depressing experience:

[Serrano] Valdenebro arrived today. He is very gloomy about the state of affairs in the Sierra, and foresees all sorts of difficulties about assembling the peasants. In a word ... the *serranos* do not want to leave their villages, and are not doing anything of any use.[19]

As La Cuadra observed, then, 'We cannot count on any aid from the Sierra, whether in terms of men, mules or anything else.'[20] The inhabitants would defend their villages from French foraging parties and punitive expeditions, and fall upon such targets as couriers and unprotected parties of civilians, but they would not take part in formal military operations. Nor was La Cuadra the only Allied officer to discover this. When Lord Blayney landed at Fuengirola in October 1810, for example, hopes that his handful of troops would be supported by the inhabitants of the neighbouring Serranía de Ronda proved unfounded, and that despite the fact that the whole area had seemingly taken arms against the French. Thus:

Previous to marching I had some conversation with Captain Miller of the Ninety-Fifth Regiment, who, with several other officers, had been latterly employed in organising the Spanish peasants. He informed me that a considerable quantity of arms and ammunition had been distributed among them, and consequently that I might expect a number to join me immediately. In this, however, I was entirely disappointed, not more than ten or twelve making their appearance.[21]

By the autumn of 1810, then, it is not difficult to find suggestions that the rising in the *serranía* should be given a more effective organisation, the need to impose a greater degree of control being strongly suggested by an ordinance submitted to the Cádiz government in November 1810 by the *alcalde mayor* of Mojocar, Ramón Somalo, which basically proposed that the de facto home guard that had been thrown up by the fighting should be levied by conscription at a rate of 10 men for every 1000 *vecinos* from among those groups of the populace not already liable to conscription (married men, for example), that the families of the men conscripted should be given public assistance, and that recruits should be formed into *tercios* under command of respected local notables.[22] But the problems touched on by this proposal did not represent an end to the problem. Meanwhile, to return to La Cuadra, he did not just voice reservations as to the military value of the insurrection, but also expressed grave doubts in respect of the direction that it was taking. 'The state of anarchy which prevails amongst the inhabitants ought to alarm us rather than give us hope.'[23] As to what he meant, this is best approached through the experiences of Ronda. As we have seen, early in March 1810 Serrano Valdenebro managed to mass a large force of peasants against the garrison. Seeing the latter's campfires dotting the hills that overlooked the town, the governor, Colonel Vinot, concluded that he was hopelessly outnumbered and pulled out without a fight. What happened next is recorded by Albert de Rocca:

The very day ... we left Ronda, the mountaineers entered it ... shouting with joy and discharging their pieces exultingly in the streets. The inhabitants of each village arrived together marching without order, and ... loaded their asses with whatever they found ... till the poor beasts were ready to sink under the weight of the booty

... The prisons were forced, and the ... criminals they contained ran instantly to take revenge on their judges and accusers. Debtors obtained receipts from their creditors by forcible means, and all the public papers were burned in order to annul the mortgages that the inhabitants of the town had upon the mountaineers.[24]

For a Spanish view, meanwhile, we have the account given by a prominent nineteenth-century antiquarian named Juan José Moreti, who was the son of an Italian printer who had come to Spain to serve the regime of Joseph Bonaparte and stayed behind in Ronda after the retreat of the French forces:

Predisposed ... against Ronda for no other reason than that it was the seat of the local law courts, the *serranos* entered the town, and, to the astonishment of all, inflicted much damage upon it, thereby sullying the sacred cause that had brought them together. As many of them had legal cases outstanding against them in the city's notorial offices ... they broke open their archives and seized all sorts of documents, which they then piled up in a heap and put to the torch ... As the town hall was located in the same square, and their object to burn everything that was written down on paper that they could lay their hands on, the municipal archive suffered the same fate.[25]

Man of order and authority as he was, Serrano Valdenebro was deeply shocked. As he later wrote, 'The atrocious conduct of the peasantry ... reduced me to tears. Much impressed with the problems of making war at the head of such unruly followers, I took my leave, and returned to my family to await whatever the future might bring.'[26] To make matters worse, meanwhile, the *serranos* could not even hold the town: On 21 March General Peyrremond arrived from Málaga at the head of a substantial column of French troops and engaged in ferocious reprisals. This brings us, of course, to a further problem with the insurrection in that, militarily speaking, the amount of damage it could inflict on the French was always very limited. A scholarly individual who had written widely on the art of war, Serrano Valdenebro was well aware of this problem. Already deeply perturbed by the wayward nature of his followers, he was not much reassured by their fighting abilities. Thus:

Although war is being waged in these mountains in the style of Viriato, flattering results cannot be expected ... The peasants are little more than unmanageable. There is little union or regularity ... in their movements. This is not to be surprised at: amongst troops who have not been fashioned by the strictest discipline they cannot be achieved ... Valiant in skirmishing, they do not understand that shock action is the chief weapon on the battlefield ... So long as troops do not realise that battles are won by the sword and bayonet, all is lost. Fire is only a chimera ... Advancing on the enemy with union and bravery ... is what brings victory.[27]

Such combatants, Serrano Valdenebro reasoned, might well be able to cause the French great difficulties, for, as he said, 'A band of patriots situated in mountains that are almost inaccessible will hold off the bravest soldiers.'[28] Warfare, though, was not just a matter of defending hill-top villages or harassing French columns as they tramped through rugged mountain passes. If victory was ever to be attained, the war would sooner or later have to be carried to the enemy, and this entailed fighting battles that the insurgents could not hope to win: 'Should the latter fall back to more accessible terrain, the picture changes ... The peasant wages a petty war ... How can this man fight in a terrain where infantry can press upon him or cavalry ride him down?'[29]

So much, then, for the 'home guard': scratch its surface, indeed, and one discovers little more than a *jacquerie* that the propertied classes had immense difficulty holding in check, little in the way of military effectiveness, and only the most limited commitment to the war effort. But what about the guerrilla bands? As we have seen, here things are a little more encouraging in that there is clear archival evidence of the worry that they caused the French high command, that the *serranía*, indeed, was a major problem for the occupying forces and as such a district that more than seems to merit its reputation as a hotbed of resistance to the Napoleonic empire. Yet here again it would appear that a considerable degree of caution is in order. On 9 November 1810, then, a captain of the regular army named Antonio García de Veas was appointed to the position of subinspector of all the guerrilla bands operating in the so-called *campo de Gibraltar* (broadly speaking, the strip of land between the Serranía de Ronda and the sea), what makes this snippet of information significant being the fact that in the Spanish army it was the task of the inspectorate not to exercise any sort of military command, but rather to ensure that peculation and other forms of malfeasance were kept to a minimum.[30] In short, it appears that the guerrilla bands were the cause of much suspicion, this impression being underlined still further if we turn to the memoirs of Thomas Bunbury, a British officer serving in the Portuguese 20th Line who was employed on a number of occasions as an 'exploring officer' in the Serranía. Thus:

> When employed on observation, collecting intelligence of the enemy, I always had with me a party of mounted Spanish guerrillas as guides. These fellows, in truth, were little else than bandits: they received provisions occasionally, but no pay. I was obliged, in consequence, to connive frequently at their misdeeds with the peasantry. I always, however, was enabled to restrain them from committing acts of violence and from extorting money. These fellows were very amusing and used to tell me stories of villainous occurrences between them and the French, but of the truth of which I could not doubt when I noticed the motley appearance of my troop, who were all clothed with the spoils of French officers. It must, in the vicinity of Cádiz, have been dangerous for the enemy to leave their lines singly or in small parties without an escort.[31]

There is at least some recognition here that the guerrillas were making an active contribution to the Allied war effort. On other occasions, however, there was much less willingness to give the bands the benefit of the doubt. Arriving in Algeciras at the head of an expeditionary division in 1811, for example, General Ballesteros gave the matter short shrift:

> The ... guerrilla bands that infested the region were one of the objects that called for my consideration. Not only had questions been raised as to the services which they had rendered, but the *pueblos* of the district were loudly complaining of their excesses ... I examined the truth of these representations in person ... Convinced

that the *partidas* existing in the district under my command were both militarily useless and prejudicial to the nation, I immediately ordered their disbandment.[32]

Very clearly, then, even the Serranía de Ronda's guerrilla bands had neither quite the impact nor the success which is often imputed to them. Why this was so can be attributed to a number of different factors. Of these, the first was beyond doubt the indissoluble link between insurrection and brigandage. Desperately poverty-stricken, prior to 1808 the whole of Andalucía had been an area notorious for banditry, and the Serranía de Ronda had been no exception in this respect. However, the Peninsular War had made matters infinitely worse. On the one hand, an economy already disrupted by two long periods of warfare with Britain (1796–1801, 1804–1808) had been subjected to further dislocation thereby severely increasing the misery of the populace, and, on the other, the imposition of conscription led to wholesale draft evasion and desertion.[33]

Even before the French came, then, the Andalusian countryside was swarming with bandits, but the catastrophe of 1810 made the situation even worse in that as the Spanish armies collapsed in the face of the enemy onslaught so they shed thousands of stragglers or *dispersos* who scattered all directions and in most instances had no option but to live by pillage, the result being the emergence of yet more gangs of brigands. Patriotic myths to the effect that at the first sight of a Frenchman every bandit in Spain instantly turned into a guerrilla, notwithstanding the swarms of outlaws had no direct interest in fighting the French, and Martinien's records suggest very strongly that the French forces did not lose a single man to guerrilla action as they pushed across Andalucía. Gradually, however, the situation began to change. In the first place, ruthless French requisitioning drove many inhabitants into such penury that it almost literally became a case of 'turn bandit or die', while, in the second, the bandits themselves had to live and therefore increasingly began to engage in actions that were indistinguishable from the actions of genuine guerrillas. Thus, although traditional targets – muleteers, shepherds, migrant labourers, local notables, isolated farmsteads – continued to suffer just as much as before, French stragglers, couriers, and convoys all began to fall prey to brigand gangs, this being, however, not so much because they were representatives of the enemy, but rather because they were easy targets; equally, if Frenchmen who fell into their hands were tortured and put to death in a variety of grisly ways, this was again not because they were Frenchmen, but rather, first, because torture had always been an inherent part of Spanish banditry, and, second, because there was usually no means of looking after prisoners.

Meanwhile, with their numbers greatly swelled – whereas the traditional *cuadrilla* numbered between 5 and 10 individuals, many had now risen to a strength of 10 times that number – and their commanders under ever greater pressure to satisfy their material wants, the gangs were also emboldened to engage in such activities as riding into isolated villages and seizing whatever resources that they offered, including, of course, the chests which held public moneys of all sorts. And, finally, confronted by punitive columns of French troops, most of which, of course, were relatively small, such gangs were not frightened of fighting back, and even of launching pre-emptive strikes against them.[34]

Clearly, then, to maintain that the bands of irregulars did no damage to the French would be very wrong. Let us not deceive ourselves, however. For all the Frenchmen that they killed – a number that was clearly rather limited – the guerrilla bands of the Serranía de Ronda were never able either to shake off their origins or to evolve into the semi-regular formations that are common in northern Spain (the most famous case here was, of course, that under the command of Francisco Espoz y Mina, but there were others including those of Juan Díaz Porlier, Francisco Longa and Juan Martín Díez). Rather than becoming an integral part of the struggle against the French, they rather circled vulture-like on its fringes.

Meanwhile, much the same was true of at least some even of the guerrilla bands sanctioned by the Patriot authorities. Some of these, certainly, gave good service – at the end of the war, for example, it was reported that the band led by Pedro Zaldivar had a very good reputation in terms of its conduct[35] – but the fact was that, even when they were not actually bandits, in which respect it should be remembered that it was all too easy for a man bent on a life of crime to hoodwink the authorities into granting him a commission, the *partidas* who ranged the *serranía* could hardly fail to teeter on the edge of banditry.

For an instructive lesson in this respect, we can do no better than turn to the case of Andrés Ortiz de Zarate. As will be remembered, this individual was not a native of the *serranía* but rather an *alicantino* who had taken refuge in Gibraltar and come to the area in company with the military mission dispatched by General Campbell. However, when the two British officers concerned had returned to their home base, Ortiz had stayed behind and sought to organise a guerrilla band. Having helped the inhabitants of Jimena beat off a French raid, he was acclaimed by them as *primer caudillo* – roughly, 'supreme leader' – of all the *serranos*. Confirmed in this role by Jacomet and now known as El Pastor, he thereafter roamed the region at the head of a band of 250 men, but seems to have done almost nothing to incommode the French, instead passing from one *pueblo* to the next in a constant search for money and supplies that edged ever closer towards outright extortion. In this situation it is possible that Serrano Valdenebro, who was angered by the manner in which Ortiz had been given a role outside the normal chain of command, may have had some responsibility: Ortiz later claimed that he had not received him the slightest logistical support which he presumably expected from Valdenebro and others. Yet, while this may have been a contributory factor, it is hard to see Ortiz as anything other than an adventurer who was bent on his own aggrandisement only and eventually became little more than a bandit chieftain, and thus it was that, when Ortiz finally travelled to Cádiz to defend himself against the growing torrent of complaint that his activities inspired, he was promptly arrested and put in prison to await trial.[36]

To conclude, then, all is not what it seems, the 'good-news story' with which we began this article proving to have grown greatly in the telling. While there was certainly a certain degree of resistance to the French at both the local and regional level, the amount of damage inflicted on the invaders was actually quite limited. Nor can this be judged to be very surprising: viewed from close up, the Serranía de Ronda seemed in the grip not of a guerrilla struggle but rather a mere orgy of pillage, the situation becoming so bad that, on occasion, the beleaguered Patriot authorities were left with no choice but to do the work of the invaders for them. In the Spain of 1810, however, such truths were inconvenient, and the result was a legend of popular heroism that has stood the test of time and looks likely to stand that of history as well.

Notes

1. *El Observador*, 24 August 1810, 134–5, Biblióteca Nacional (henceforth BN), Colección Gómez Imaz, R60632.
2. For a discussion of these apologia in particular, cf. Martín de Molina, 'La guerrilla en la Serranía' vista por cinco de sus protagonistas." In *Cuadernos del Bicentenario*, 10 (2010): 191–207.
3. E.g. Henegan, *Seven Years' Campaigning*, 2: 166–9.
4. E.g. Haythornthwaite, *In the Peninsula with a French Hussar*, 156.
5. E.g. Martínez Laínez, *Como lobos hambrientos*, 394.
6. For a description of the Serranía de Ronda on the eve of French occupation, cf. Jacob, *Travels in the South of Spain*, 323–50.
7. Díaz Torrejón, *Guerrilla, contra-guerrilla*, Vol. 3, 3: 164–8.
8. Martín de Molina, *Gaucín*, 114–21; cf. also Serrano Valdenebro, *Manifiesto*, 12–13. BN. CGI. R61110.
9. Gutiérrez Tellez, *Biografía de D. José Serrano Valdenebro*, 80–1.
10. J. Serrano Valdenebro to A. Jacomet, 5 May 1810, Archivo Histórico Nacional (henceforth AHN), Sección Diversos-Colecciones, *legajo* 94, No. 1; J. Serrano Valdenebro to A. Jacomet, 2 June 1810, ibid. Haythornthwaite, *In the Peninsula with a French Hussar*, 134–5; Henegan, *Seven Years' Campaigning*, 1: 199–200.
11. Cf. Lavaux, *Mémoires de campagne*, 153–6. Proclamation of B. de Aranza, 8 May 1810, AHN, Sección de Estado, *legajo* 2994. In this action, Algodonales does indeed seem to have been treated very harshly: in 1814 the then parish priest compiled a list of no fewer than 239 of the inhabitants who had perished in the attack, including a child just three days old; cf. Díaz Torrejón, *Guerrilla*, 3: 163.
12. Díaz Torrejón, *Guerrilla*, 3: 225–40 *passim*.
13. Henegan, *Seven Years' Campaigning*, 1: 171–2.
14. Haythornthwaite, *In the Peninsula with a French Hussar*, 163.
15. J.D. Soult to A. Berthier, 8 February 1811, Service Historique de la Défence, Section de Terre, T. C8–147, ff. 32–5.
16. Cf. Martinien, *Tableaux*.
17. I owe these figures to my good friend and colleague, Jorge Planas Campos. It should be noted, however, that an absence of officer casualties does not necessarily equate to an absence of any officers whatsoever. From documents in Soult's correspondence,

then, we know that the operations against Algodonales cost 45 dead and 108 wounded; cf. J.D. Soult to A. Berthier, 8 May 1810, Service Historique de la Défence, Section de Terre, C-146, ff. 245–7.

18. F.J. de Castaños to J. Blake, 28 March 1810, Instituto de Historia y Cultura Militar, Colección General Blake, *legajo* 3, *carpeta* 25.
19. A. de la Cuadra to F. Abadia, 13 June 1810, ibid.
20. Ibid.
21. Blayney, *Narrative of a Forced Journey*, 13–14. NB According to Blayney the presence of Miller and the other officers, of whom the most prominent appear to have been two lieutenant colonels named Basset and Warrington, was the work of the Duque del Infantado, the latter having prevailed on the British government to send a number of British officers to organise the *serranos*. Ibid., 26–7.
22. 'Reglamento para la formación de cuerpos patríotas que hagan el servicio de guerrilla en el Reino de Granada interín su ocupación por el enemigo', AHN, Sección de Diversos-Colecciones, *legajo* 124, No. 28.
23. A. de la Cuadra to F. Abadia, 13 June 1810, Instituto de Historia y Cultura Militar, Colección General Blake *legajo* 3, *carpeta* 25.
24. Haythornthwaite, *In the Peninsula with a French Hussar*, 150–1.
25. Moreti, *Historia de L.M.N.Y.M.L Ciudad de Ronda*, 598–9. It was not just legal documents that were burned: also consigned to the flames was a large quantity of *bulas de cruzada* that were seized in the house of Antonio Gómez Barroso, the latter being the official responsible for their sale; cf. Conde de Montarco to J.M. Sotelo, 12 February 1812, AHN, Sección de Consejos, *libro* 1745, f. 104.
26. Serrano Valdenebro, *Manifiesto*, 18.
27. J. Serrano Valdenebro to J. M. Carvajal, 4 April 1811, citing *Diario de Algeciras*, 24 April 1811, pp. 357–61, Hemeróteca Municipal de Madrid, AH227. Viriato – more commonly, Viriatus – was a leader of resistance to the Roman Empire among the tribes of modern-day Portugal.
28. Ibid.
29. Ibid.
30. Díaz Torrejón, *Guerrilla*, 3: 233.
31. Bunbury, *Reminiscences*, Vol. 1: 97–8.
32. Ballesteros, *Respetuosos descargos*, 21–2.
33. For an extended discussion of both the social structure of *antiguo régimen* Andalucía and the experience and impact of the Peninsular War, cf. Esdaile, *Outpost of Empire*, 41–132.
34. Cf. ibid., 323–4. All this said, it is important to note that by no means all gangs of bandits ended up fighting the French: on the contrary, at least some offered their services to the invader and were transformed into auxiliary counter-guerrilla units, while it is also clear that many individual *malhechores* gave up their lives of crime to join such *cuerpos francos*; cf. ibid., 288–90.
35. Capitanía General de Andalucía: Noticia de las partidas de guerrilla o cuerpos francos que se formaron en los Reinos de Sevilla, Córdoba y Jaén en los años de 1809, 1810, 1811, 1812 y 1813 en la pasada campaña, 1 August 1817, AHN. Diversos-Colecciones, 124, No. 50.
36. Díaz Torrejón, *Guerrilla*, 3: 26–31. There is another side to this story. Thus, in a series of manifestos, Ortiz denied being guilty of any wrongdoing and was eventually cleared of all the charges that had been laid against him. However, given the atmosphere that reigned in Cádiz at the time of his trial, an acquittal meant nothing, for by this time Ortiz had become a convinced supporter of political and social reform, and was therefore being championed by the clique of radicals responsible for pushing through the constitution of 1812. This did not prevent both Southey and

Schepeler from supporting Ortiz's innocence and making a variety of more-or-less exaggerated claims in his favour, but his case remains dubious at best, the fact being that, as Díaz Torrejón suggests, there is just too much evidence against him for everything to be explained away in terms of conspiracy on the part of rivals and political enemies. Meanwhile, it is no credit to Ortiz at all that he seems to have been behind the disastrous attempt to disembark a force of British troops at Fuengirola that led to the capture of Lord Blayney.

The German wars of liberation 1807–1815: The restrained insurgency

Martin Rink

Zentrum für Militärgeschichte und Sozialwissenschaften der Bundeswehr, Potsdam, Germany

In the Age of Napoleon, 'small wars' and 'revolutionary war' were closely connected. There were, however, different strands of this phenomenon: speaking professionally, conservative officers condemned small wars as an irregular regression to previous less disciplined forms of warfare. The Prussian state continually tried to discipline and regulate spontaneous risings. Yet the irregular character of small wars offered the opportunities for a less complex way of fighting, thus enabling the arming of the 'people' to fight. Individual undertakings, such as Ferdinand von Schill's doomed campaign in 1809, were designed to spark off a general popular uprising. But they were cheered by many and supported by few. Meanwhile, Neidhardt von Gneisenau conceived guerrilla-style Landsturm home-defence forces, which were designed for an irregular people's war. These concepts were put into practice in the 'war of freedom' – or 'war of liberation' – in 1813. Eventually both the mobilisation and the tactics remained regular, however, despite the emphatic appeal to a national 'people's war'.

The Prussian reforms and 'revolutionary war'

'In the beginning was Napoleon.'[1] With this much-repeated quotation, historian Thomas Nipperdey began his history of the German nation in the nineteenth century. It is precisely because of this 'national' redefinition that the quarter of a century between 1789 and 1815 appears as an 'axial age' or 'saddle period'.[2] The rise of the 'nation' as a sociopolitical force was combined with the transformation of 'small wars' (*petite guerre, kleiner Krieg*) from special operations to popular uprisings or 'people's war' (*Volkskrieg*).[3] By the same token, small wars in the Napoleonic period mirror this transformation. In Prussia, they mark a political/military turning point. The defeat of the Prussian army at Jena and Auerstedt on 14 October 1806 is a classic example of the failure of the *ancien régime* and its armies when pitted against Napoleon. While before 1806, the old Prussian army had been aware of innovations,[4] it was not until the Prussian catastrophe that a 'people's war' could be contemplated. In the winter of 1806–1807 there were already minor yet highly effective stirrings aiming at the 'liberation' of Prussia or its lost territories. The kingdom of Westphalia in particular, half of which comprised former Prussian territories, was repeatedly the target of patriotic Prussians' plans, as in the attempted uprising of 1809 and the 'war of liberation' of 1813.

To the influential military historian Werner Hahlweg, 'small wars' and 'revolutionary war' were closely connected.[5] But his colleague Johannes Kunisch had already emphasised for the pre-revolutionary eighteenth century that this form of conflict reflected prevailing 'political and social conditions'.[6] Certainly, this is true in view of the 'revolutionary' warfare practised by the French armies after 1792–1793. Given the complexities of the subject, it is necessary to consider the tactical, political, and semantic developments of these phenomena: the *practical side* of this form of conflict always varied considerably. As those practising it were so heterogeneous, is it actually possible to speak of *one* form of conflict in the singular? Even the notional concept of small wars was subject to continuous change, as the present collection of essays demonstrates.[7] It therefore makes sense to divide this examination of small wars during the Prussian reform period into three strands: first, the *tactical concept* of small wars in Prussia from the eighteenth century until the wars of 1813–1815; secondly, the *concepts of a people's war* in the Prussian reform period; and, thirdly, the *military organisation* of the forces that were intended to conduct small wars, notably in the spring of 1813.

The earliest 'classical' literature on small wars, strongly influenced by the War of the Austrian Succession (1740–1748) formed the point of reference for the tactical concept of small wars during the Prussian reform period. In late 1740, young Prussian King Frederick's assault on Silesia started his career of a *roi-connétable* – the king who is his own chief general – par excellence: At the latest with his smashing victory at Hohenfriedberg/Striegau on 4 June 1745 against superior Austrian-Saxon forces, Frederick earned the sobriquet 'the Great'.[8] Yet the Silesian Wars also led to irregular forces from the region of the Habsburg military border being channelled in considerable numbers towards Central Europe. Thomas de Grandmaison, the defining French expert on the '*petite guerre*', wrote of a 'torrent of light and irregular forces of the Queen of Hungary who inundated Bohemia, Bavaria and Alsace under circumstances in which France found itself bereft of forces of a similar kind'.[9] Warfare, therefore, assumed two forms: major battles, on the one hand, and small wars, on the other. With Frederick in mind at least in most German-speaking areas, open battle remained the only benchmark for military greatness. Prussian forces had to overcome this cultural taboo in order to adapt to the latter 'un-Prussian' form of conflict. They found this difficult, but ultimately did so successfully.

In the mid eighteenth century, light forces were the main identifying feature of small wars, although this would not have said anything substantial about the tactics used in them. While there was hardly anything 'new' either tactically or with regard to the actors in this connection,[10] small wars were something 'novel'. From the middle of the eighteenth century onward, publications on this developed into a proper literature genre. Seasoned practitioners would initially write adventure stories about their experiences. This led, in the middle of the century, to handbooks based on their own war experiences that were systematised, though, into empirical rules. Later, tactical guidelines and training instructions then evolved. These works first appeared primarily in French, and later also in German, while English-language works were initially not very widespread.[11] Towards the end of the century, officers who underwent their military socialisation in the decades of peace of the 1770s and 1780s helped to promote this conceptualisation of small wars. Lacking first-hand experience, they drew on the literature. The best example of such a systematic and didactically prepared evaluation of 'extraneous' war experiences is provided by the Hanoverian military author Gerhard Scharnhorst.[12]

The lessons learned from the American War of Independence (1776–1783) were also gradually assimilated to the extent that sniping became an instrument of small wars. This happened even *before* the appearance of the French '*tirailleurs*' (German: '*Scharfschützen*', literally 'sharp shooters') of the revolutionary wars. In the 1790s, these conflicts sparked a lively debate in German-speaking military discussion forums and publications about the degree to which they embodied a revolutionarily new image of war.[13] The outbreak of the revolutionary wars against the 'New French' of revolutionary France appeared to increase the interest in small wars and their derivatives, especially the use of snipers. From the 1790s, these earlier works were, as a result, re-edited or served as a point of orientation for the numerous newly appearing publications.[14] Nevertheless, in the tactical sense, there was no 'revolutionary' break with the 'classical' small war of irregular detachments, '*kleiner Krieg*' in German. The publication in 1809 of an adapted translation and of a new edition of a classic that had initially appeared in French in 1756 bears witness to the interest in an apparently fashionable topic as well as to the timeless continuities. Entirely in the spirit of the new age, however, the publisher of this relevant handbook emphasised that everything was now 'small wars on a large scale'.[15]

In a military-organisational respect, those involved in small wars were subject to a continuous 'insourcing' process. This 'nationalisation' of the armed forces, and their regularisation and absorption by the (emerging) state had largely been completed in the Prussian army after the Seven Years War. Conservative Prussian officers, proud of this regularisation and disciplining, therefore disapproved of the Prussian reformers' plans to raise militias or arm the population. This also applied to officers who were experts in the field of small war tactics, like Hans David Ludwig von Yorck, who in the debates would become an opponent of August Neidhardt von Gneisenau. As a result, at the beginning of the nineteenth century, and particularly during the Prussian reform period, small wars would be 'nationalised' in such a way that the partisan units were not seen simply as acting on behalf of the state (or of the ruler), but as defenders of the 'nation'.[16] Small wars became associated with irregular innovation or were condemned as regression, and were followed by periodic attempts to regularise the partisan units.[17]

Concepts of a 'people's war' in the Prussian reform period, 1807–1813

From the late eighteenth century, the proto-national movement in Germany was dominated by the educated middle classes. The 'reading revolution' of that time expanded the horizons of thought and communication, thus uniting German-speakers beyond their territorial affiliations. New elites defined themselves against the establishment through the ideal of education and meritocracy. The representatives of the *Sturm und Drang* period idealised the 'wild youths'. Friedrich Schiller, a regimental surgeon from Württemberg who had deserted his unit, in his first work *The Robbers* (1781) condemned the 'ink-blotching castrati century'.[18] Around the same time, an ideal image of the 'noble savage' was able to establish itself through Jean-Jacques Rousseau's glorification of nature; initially, this, too, remained an avant-garde phenomenon.[19] The promotion of the ideal of nature during the revolutionary period, however, enabled light troops, who equally appeared as wild, uncivilised, and 'natural', to provide the stuff of which later on heroes could be made. It was thus possible in the early Romantic period for a new cultural and military ideal to emerge which originally had little in common either with the light troops, or with the intellectual circles, or with the protagonists of the national movement. This provides a link to those who later rose up against Napoleon – either as 'patriot partisans' or 'insurgent bandits', depending on one's point of view. The wild and romantic type of freedom fighter would be personified by Ferdinand von Schill in 1807, and conjured up in the memoranda of Gneisenau and Carl von Clausewitz from 1808 to 1812.[20] This paradigm was followed by the isolated attempt at an uprising by that same Schill in May 1809 and perpetuated itself in the proclamations for a 'war of liberation' in the spring of 1813.

Also in the late eighteenth century, a form of military literature had established itself which, imbued with the enlightened ideal of the educated officer, was nevertheless pragmatically oriented. Scharnhorst offers the best example of this development.[21] His name is associated both with the scientisation of military affairs in the Age of Enlightenment and with the application of its findings in the Prussian reforms. A peasant-boy by origin, he was admitted as a cadet to the military school of William, Count of Schaumburg-Lippe.[22] Scharnhorst's indefatigable military publishing activity while still in Hanoverian service as an editor and author from 1785 onward testify to this military education. Scharnhorst's *Military Handbook*, which he published in 1793, was essentially a textbook on small wars. The major part of this field manual for the officer in charge of a detachment dealt with reconnaissance and skirmishing duties, much the same way as did previous books on small wars. The author's didactic skills made the *Military Handbook* an excellent training manual, all the more so as his style departed elegantly from the dry style of the old army's drill manuals.[23] It continued to have a strong influence in the campaigns after 1813. This demonstrates that some elements of pre-revolutionary tactics were well suited to the Napoleonic era.[24] Scharnhorst, after entering into service with the Prussian army and acquiring the mark of nobility, first as a teacher at the military service school, then as de facto war minister, reformed Prussia's military, fusing old and new elements. Compulsory military service was central. Tactical elements of Frederician warfare were merged with Napoleonic ones. 'Small war' remained only one element among many, but had a considerable influence on the mood of the reform period. The emphasis on 'small war' in the reform period had a systemic cause: it aimed to simplify tactics. In the age of Absolutism, military and organisational procedures had become excessively complex. In the 1790s, this was derided as parade antics in the enlightened literature,[25] just as the French ridiculed the 'lace warfare' of the *ancien régime*. French skirmishers, by comparison, seemed to point the way to the future of tactics. Admittedly the French revolutionary troops were victorious not so much due to their enthusiasm but because of their numbers, nor were they always victorious.[26] Later on, however, their way of fighting contributed to making the Napoleonic army the most effective military machine around 1805–1806.

The French Revolution was the starting point and remained the point of reference for the liberation movements of the 'modern era'. The Prussian reformers would make reference to the revolutionary paradigm of the 'people's war' – in marked contrast to their conservative opponents and to the displeasure of their king, Frederick William III, who played off the two parties against each other. Away from the military-organisational and tactical aspects, the legitimisation of war-related violence changed with the French Revolution. As a result of new tactics and new actors there was, in many respects, a break with the traditional *ius in bello*, in other words with the conventions applicable in war. Admittedly they had been disregarded often enough under the *ancien régime*, and yet many a chivalric war practice had been retained, such as the release of officers held as prisoners of war after they had given their word of honour.[27] Over the previous centuries, the right to go to war had gradually been monopolised by the state. As the legitimisation of war now related to the 'people' or the 'nation', it was now possible for violent non-state actors to claim they were conducting legitimate (albeit, in the formal legal sense, not necessarily lawful) warfare. For this reason the changes in the wake of the French Revolution concerned not only the tactical but mainly the strategic level. The causes and purposes of war changed. And along with them, the function of – tactically largely unvarying (and largely regulated) – small wars changed.

In Germany, any new way of arming 'the people' remained closely regulated. In the 1790s, militias were raised for home defence only in some areas on the Upper Rhine. Those, however, were rather pre-modern, now reactivated, forms of military service.[28] Discussion forums also sprang up in which the new type of warfare was discussed, one example being the *Militärische Gesellschaft* (Military

Society) founded in 1801 in Berlin on the initiative of Scharnhorst. Even though mainly tactical and practical aspects were discussed, it was obvious to their members that politics and warfare were interlinked for the French revolutionary soldiers.[29] Aspects of 'people's war', although not yet described as such, manifested themselves in traditional forms of resistance, where there was opposition to taxation and conscription and especially where there had been a change of ruler in the course of Germany's territorial reorganisation between the years 1801 and 1807. Consequently, the great commotions in the wake of the Revolution also saw a revival of traditional brigandry. This led to similar forms of violence as those encountered in Spain or southern Italy at the time – albeit with a lower intensity.[30] In Germany, too, the radical upheaval engendered unrest and revolts, sparked by the measures and impositions of the new order. In the north German model state of Westphalia, these seeds of unrest were intensified by former old Hessian or Prussian soldiers, deserters and conscription evaders, with old loyalties exerting an influence just as much as plain poverty.[31] This mix of motivations formed a backdrop for the 'patriots'' plans for uprisings between 1807 and 1809. The Prussian defeat of 1806 and subsequent occupation and changes in state structures and boundaries in northern Germany shook the old order. In the presence of French occupation forces, earlier national stereotypes such as the image of the 'frivolous Frenchman' or the 'stuffy German' were intensified. During the 'French period', the French were the 'other' against which German nationalist sentiments could crystallise.

Only a few weeks after the Prussian defeat at Jena and Auerstedt, Ferdinand von Schill rose to become the prototype of a protagonist in future people's wars. While this uprising remained localised, it happened 15 months before Spain set a European-wide model of a popular revolt. The previously unknown lieutenant, who had been wounded near Auerstedt, managed to make his way to the East Pomeranian garrison of Kolberg. There, on his own initiative and initially without any authorisation – certainly not from the conservative fortress commander – Schill formed a small volunteer corps. It recruited other Prussian soldiers also streaming back from the battlefields of Jena and Auerstedt, as well as peasants with improvised weaponry. Irrespective of the limited military value of his actions, Schill set an attention-grabbing example, demonstrating that Prussia's resistance was unbroken. The townsfolk of Kolberg illustrated that previously untapped resources among the population could be mobilised. As in other parts of Europe, though, the Kolbergers followed old example, for instance, antecedents of the Seven Years War. The eighteenth-century professionalisation of the Prussian military, with its contempt for traditional local militias and for small wars, was thus called into question. The very elements of warfare that conservative officers saw as unprofessional retrogression now appeared to be the way forward. The deeds of Schill and the Kolbergers received effusive press acclaim. An anonymous article published in 1808 enthused: 'Everywhere we heard laudable things about brave Schill; people speak his name everywhere with a look of joy in their eyes. "Indeed, if that man had had more power, a larger sphere of influence, then what could have been saved surely would have been saved!", was to be heard on all sides, and I think with good reason.'[32] On the wings of popular enthusiasm, Schill rose up at breathtaking speed to become a popular star – decorated by the king, honoured with favours by the queen, cheered by the population, and depicted on copper engravings, tobacco and snuff boxes, and also as a cake decoration figure.[33] The period between November 1806 and February 1807, when old Prussia collapsed, was the formative phase for the Schill myth. It was around the same time that, in Prussia and throughout northern Germany, small wars were popularised as 'people's wars'. Partisan units, formerly seen as suspect in view of many war atrocities for which they had been blamed, were thus transformed into the people's liberators. However, the recasting of the German mindset remained a blurred and complex process. After the defeat, King Frederick William III fled from his palaces in Potsdam and Berlin (capital of the kingdom of Brandenburg-Prussia) to Königsberg, the capital of the duchy of (East) Prussia, his most remote province. This left a power vacuum in Brandenburg, the centre of the Prussian monarchy. And this in turn offered a window of opportunity for decentralised powers to proclaim resistance – much as in Spain after the events of May 1808. And there, it was not until the turn of 1808–1809 that the Spanish 'patriots' of the ruling juntas drew up a concept of an irregular people's war. Despite the poor flow of communication between Spain and Central Europe, German newspapers showed awareness of Napoleon's problems with the Spanish insurgency. This, in turn, had an impact on the plans being made in Brandenburg, northern Germany and in Austria, accompanied by patriotic propaganda that glorified people's war.[34] At the same time the Peninsular War demonstrated how the French would have dealt with German insurgents, namely with brutal force.[35] Nevertheless, the 'nationally' inspired decrees of the Spanish patriots shaped the perception of the Prussian reformers. German men of letters translated them or formulated pamphlets based on a similar pattern.[36] The *guerrillero* had become the archetype of the liberator of oppressed peoples.

In April 1809, Austria used this power vacuum by launching a war against France, claiming to act on behalf of the whole of Germany.[37] But the German-speaking peoples were by no means united in their views. Two days after the outbreak of the war, an anti-Bavarian uprising began in Tyrol, which had only recently been annexed by Bavaria, France's ally. Here, too, the boundary between internal and external conflict remained blurred, with the fighting taking place *guerrilla*-style, and the rebels also referring to earlier Tyrolean liberties, which the Bavarian reform-oriented state had taken away from them. In particular, new taxes and conscription fuelled the resistance.

It was in this context that the Prussian reformers Karl Freiherr vom und zum Stein, Scharnhorst, and Gneisenau began to prepare an anti-French popular uprising. On 11 August 1808, Stein presented a memorandum to the King of Prussia, urging him to side with Austria.[38] Scharnhorst drew up detailed plans for the implementation.[39] The same month, Gneisenau outlined further plans in a memorandum. His subsequent memorandum of 1811, Clausewitz's 'memorandum of confession' of 1812, and the '*Landsturm*' edict proclaimed in April 1813 contained clarifications which closely paralleled the Spanish appeal of April 1809.[40] Gneisenau explicitly referred to 'Spain's noble example' while urging a popular uprising in the ex-Prussian territories in the kingdom of Westphalia.

The possibility of such an insurgency arising in Prussia was feared not only by the conservatives. For this reason, the Prussian reformers denounced them as the 'French party'.[41] But Frederick William III forbade the planned guerrilla-style uprising. Whereas in 1808 and 1809 he was attracted by his subordinates' plans, in 1811, he denounced the idea of a 'revolutionary people's war [*Volkskrieg*] which rushes everything into confusion', it would be 'bedlam' ['*Tollhäuseleien*'].[42] Stein, as the chief Prussian reformer, had been dismissed from office in November 1808 at the request of the French Emperor.

Schill himself illustrated the opportunities and limitations offered by a people's war. In early December 1808, he and his unit, upgraded to a regular regiment, triumphantly entered his new garrison of Berlin and instantly began new planning for a general uprising.[43] In April 1809, an attempt by (former) Prussian officers to instigate an insurrection in the (former Prussian, now Westphalian) Altmark region failed. The rebels dispersed when no support came from Prussia. Nevertheless, Schill continued to plot together with other 'patriots'. Together with Adolf von Lützow, the later volunteer corps leader of 1813, and others, he prepared an appeal to the population of Westphalia to rise up. When this was leaked, he decided to take action. On 28 April 1809, he marched his unit to the gates of Berlin and addressed his soldiers, asking them if they were prepared to follow the example of Spain and the Tyrol, to 'take vengeance on the hated enemy for the humiliation of the fatherland'.[44] Again, an uprising failed to materialise. Only a few veterans and civilians supported this call to arms of a regiment acting on its own initiative. Obviously, despite the somewhat undermined authority of the king, loyalty towards him – and the old insistence on law and order – still prevailed among the officer corps.

With this act, Schill had turned his troops from a regular unit back into a volunteer corps. From a legal perspective, there was no longer any difference between Schill's men and a band of outlaws.[45] In Westphalia, Napoleon's brother, King Jérôme, had proclaimed Schill an outlaw. Nevertheless, a light infantry battalion, which had previously been under Schill's command, left Berlin on its own initiative in order to join him. This was open desertion. The applause and enthusiasm in the gazettes and taverns of the Prussian capital stood in clear contrast to the reluctance of the population to turn into insurgents and outlaws themselves.

Schill's undertaking illustrates the risks involved in a people's war: a spontaneous, unauthorised campaign without the approval of the ruler might be a preliminary step towards anarchy. To the great relief of the conservatives, it did not come to this. After skirting the kingdom of Westphalia to the east, Schill headed for the Baltic, to the port of Stralsund. On 25 May, he entered the town following bloody fighting and even rioting against the French occupation forces.

He raised *Landsturm* (home defence) units from the local population, but on 31 May, Dutch and Danish forces, allied with France, won the town back in a short and fierce engagement. Schill was killed in action.[46] His campaign, designed to spark off a general popular uprising, ended as a foray, which was cheered by many and supported by few. In his refuge in remote Königsberg, Frederick William III dissociated himself from Schill, but had not stopped him. Nevertheless, the volunteer corps held out for a whole month without any external support, having annihilated a Westphalian regiment in open battle and, by doing so, having drawn attention to the military weakness of this Napoleonic model state.

Schill was not the only one to fight on his own initiative. He was followed, a month after his campaign, by the 'Black Duke', Frederick-William of Brunswick-Oels, who was the heir apparent to the duchy of Brunswick. Deprived of his lands, which had been incorporated into the kingdom of Westphalia, he formed a volunteer corps in Bohemia and fought as an 'ally' of Austria. After a campaign similar to that undertaken by Schill, he reached Brunswick in early August. Shortly thereafter, following a bitter but indecisive engagement, he was left with no option but to take the fast route to the North Sea coast and embark on British ships that came to his rescue.[47] The uprising of 1809 had petered out in courageous yet isolated acts. Nonetheless these were only the more significant resistance movements among many decentralised acts of unrest directed against the occupying power. Collectively, they posed a latent threat to the traditional order of the local elites. In 1809, examples of a peoples' war had been set. It had failed for the time being due to a lack of coordination.

After Stein's dismissal in December 1808, planning for a popular uprising within the Prussian government ceased. Only a small minority of men continued to write about such an option, among them Gneisenau. His *Landsturm* concept of 1811 envisioned total resistance in the manner of a nationwide people's war – now on a strategic scale. Gneisenau had recognised that the German princes' commitment to law and order buttressed the rule of Napoleon. As in the wars of the *ancien régime*, only far more efficiently, Napoleon's troops drained the resources of the countries they occupied – with the help of obedient local bureaucrats. Gneisenau's concept was at odds with the maintenance of such public order. Tactically, this concept contained characteristics of 'small war'. Negating the conservatives' creed, Gneisenau asserted that 'the German soldier had superiority' in this method of fighting. Because of their inferiority in open battle, the militias should fight only in small, surprise engagements. Gneisenau stated that 'moral principle', in other words 'enthusiasm' and a 'sense of national identity', was needed for his concept to be successful.[48] Gneisenau was unable to persuade his king of this who continued on his course of risk minimisation.[49] As the *guerrilla* had created broad zones of anarchy in Spain, which from 1814 posed major problems for the Bourbon king on his return,[50] the Prussian king's 'policy of inactivity' can also appear 'extremely wise' in retrospect.[51] Nevertheless, through his *Landsturm* concept, Gneisenau had designed a model for people's war which could be put into practice at the next opportunity. That opportunity presented itself after Napoleon's *Grande Armée* had perished in Russia.

Forms of mobilisation in the Prussia of the reform period

The way for the Prussian-German War of Liberation of 1813 was paved by further unauthorised initiatives on the part of Prussian officers. On 30 December 1812 the conservative Yorck, who under Marshal Jacques Macdonald had led his Prussian auxiliary corps into the Baltic territories, declared his corps' neutrality towards the Russian army. He thus presented his monarch with a fait accompli. Stein, in exile since 1808, was now, on behalf of the Russian occupying power, granted the administration of the province of East Prussia where the assembly of estates raised a *Landwehr* militia force. Indeed, this was a 'war of liberation' – though one initiated by Russia. Frederick William III's monopoly on the use of force slipped from his hands. Although force commanders and estates acted in his name, they did so autonomously; although they informed the king, they did not wait for his approval. This bore old pre-absolutist or new revolutionary traits, depending on one's perspective. Indeed, the monarch himself had pondered the risks and opportunities of changing sides in late 1812, and was on the brink of breaking his unequal alliance with Napoleon. Given his subordinates' interfering in his prerogative (the decision about war and peace), he now risked losing face and honour if he did not put himself at the head of this movement.[52] Despite the danger of a power vacuum in Prussia, and despite his dithering, in contrast to the situation in Spain, Frederick William III was still the recognised ruler of Prussia. To regain his authority, he mobilised the state's traditional instruments and eventually exercised 'national' leadership – be it Prussian or German.

On 16 March 1813, Frederick William finally declared war on France and, the day after, made his famous appeals *An Mein Volk* ('To my people') and *An Mein Kriegsheer* ('To my army'). These recognised the fundamental change in the official conception of war. The king now whipped up national feelings with a religious fervour. The wording was ambivalent with regard to the people who were to be liberated – only Prussians or all Germans? The first appeal was addressed to both the 'nation' and the 'fatherland'. All Prussians and other Germans who felt this was addressed to them were thus included.[53] Between February and April 1813, the Prussians were mobilised. Universal conscription ensued, in five steps. First, the line units were increased in number, with all previous exemptions from military service being revoked on 9 February. Secondly, on 17 March 1813, edicts were issued to raise a *Landwehr* force. This militia force was to be raised on a district-by-district, i.e. a decentralised, basis. The *Landwehr* units would have to provide their own clothes and, where possible, their own weapons. In any regular engagement they were to support the line troops, yet elements of small war tactics and of a popular uprising could be seen. A cloth cap bearing a white cross and the inscription 'With God for King and

Fatherland' signalled the national people's war. In the battles near Grossbeeren, Hagelberg, and Dennewitz in late August and early September 1813, it was mainly *Landwehr* units who prevented the French/Rhenish Confederation troops from advancing on Berlin. Thus they – and along with them the conscript soldiers and the idea of conscription – made themselves immortal in Prussian-German mythology.[54]

Besides its core, comprising line and *Landwehr* elements, the contingent of 'the' people was supplemented by further formations. Thirdly, on 8 February 1813, a proclamation on the formation of voluntary light infantry detachments was issued.[55] These formations were to operate like earlier volunteer corps, but were now intended as an incentive to mobilise the middle classes. The military draft exceptions that had previously applied to them were cancelled the very next day – good reason for them to take the opportunity to enlist rather as volunteers in special units than as common soldiers. Fourthly, on 9 February, Lützow and some fellow officers requested the institution of such volunteer corps of non-Prussian Germans, a request that was granted by the king nine days later. Among these, the Lützower would become the most famous. Unlike the light infantry detachments, these corps were not to be attached to the regular regiments. They were, instead, to operate on their own in the enemy's hinterland, especially in the former Prussian territories, to win over the population for the patriotic cause. Fifthly, the concepts of a people's war reached their declaratory climax through the *Landsturm* edict of 21 April 1813. This edict for raising a local home-defence militia reflected the spirit of Gneisenau's earlier *guerrilla* concepts. In Prussia, however, and wherever the troops of the coalition army advanced in the autumn of 1813, reality presented the same picture. A *Landsturm* was proclaimed with an enthusiastic rhetoric to defend the fatherland, fighting in the manner of the *guerrilleros*. Each time, however, these formations were deployed as a local constabulary force – watched over closely and suspiciously by bureaucracy and the establishment. At the end of July, the edicts concerning *Landsturm* units were toned down, losing any reference to a spontaneous national uprising. On 4 March 1814, the king issued the order to suspend all *Landsturm* exercises.[56]

In fact, this 'universal conscription' turned out to be a target-group-oriented project. But indeed, it led to a consistent growth of numbers in the army. The strength of the Prussian army at the beginning of the spring campaign of 1813 was only 65,000 men.[57] By the beginning of the autumn campaign, the *Landwehr* forces had increased in strength to 109,000 foot and 11,000 cavalry soldiers. They thus accounted for roughly half of the 279,000 mobilised Prussians in total and formed the very core of the Prussian army reform. That was about a fifth of the Prussian male population able to bear arms.[58] The modes of mobilisation of the *ancien régime* and the 'people's war' were combined organisationally. It was not, however, a 'revolutionary war'.

The mobilisation of manpower in Prussia was supplemented by that of 'moral forces' – as in revolutionary France and anti-Napoleonic Spain. A wildly romantic notion of 'people's war' emerged that simply did not fit the reality.

When the freshly mobilised troops were undergoing their drills, the reality proved to be a mixture of awkwardness, standoffishness, and reservations and conflicts among the different social classes that came together. Inexperienced stressed *Landsturm* officers, such as young lawyers or elderly dignitaries, commanded at times lukewarm troops recruited from among 'the people'.[59] Nevertheless, the initial elite phenomenon of patriotism spread in March 1813, reflected in symbols that have since been associated with the Prussian army reforms. In his memorandum of August 1811, Gneisenau proposed the creation of a 'Prussian-German decoration' that could be earned by any soldier, irrespective of rank, solely on the basis of merit.[60] Two years later, on 10 March 1813, Frederick William III finally espoused this idea and instituted the Iron Cross. It was the most manifest expression of the changes in Prussia after 1806,[61] a step towards equality among the people.

The outcome of the wars of 1813–1814 took two forms, reflected by the alternative terms 'war of freedom' and 'war of liberation'. The former emphasised the spontaneous national uprising of enthusiastic defenders of the fatherland as the most important reason for the overthrow of the Corsican 'tyrant'. This is illustrated by the appeals disseminated by the Russian/German (exile) detachments in northern Germany and in Saxony in 1813, and also by the romantic death-wish songs of Theodor Körner and the aggressive propaganda of Ernst Moritz Arndt. The latter, through his publications, propagated a people's war that would practically eliminate the 'difference between warrior and citizen'. The sanctification of war for the fatherland was combined with patriotic and revolutionary rhetoric in popular, sermon-style language: 'He who fights tyrants is a holy man, and he who subdues haughtiness serves God.... But cursed be the name of him who fights for the tyrant and draws the murdering sword against justice.'[62] Theodor Körner also demanded universal military service for all male – and thus manful – Germans while denouncing the 'boyish squirts by the fireside, among the maids'.[63]

In the end, the much-praised alliance between people and crown was one-sided. The extensive mobilisation was followed by measures to channel the popular forces back into bureaucratically controllable state structures. Nothing illustrates this more clearly than the revocation of the *Landsturm* concept. According to the Prussian conscription law issued on 3 September 1814, *Landwehr* and *Landsturm* units formed a reserve for a standing conscript army organised as line-type units. This was peoples' war provided that there be an orderly mobilisation.[64] A real, spontaneous, general Prussian people's war in its irregular form did not take place – with the exception of some very few brilliant individual feats and many plans and concepts. Underlying these were the Prussian reformers' concepts of a people's war, from Scharnhorst's plans for a broader mobilisation of the population, to the guerrilla-style *Landsturm* corresponding to Gneisenau's memoranda. Clausewitz, then only a junior staff officer, drafted similar plans at the time on which he drew when he wrote his *On*

War in the 1820s,[65] but unlike Gneisenau, Scharnhorst, and Stein, he would have no influence in his own lifetime.

The restrained insurgency: The Prussian way of peoples' war

If people's war had its beginnings in the era of revolutions, so does the term itself in all its ambiguity. Its complex semantics contributed to the political conflicts of the nineteenth century. A quarter of a century of revolution and war led to 'mercenary' becoming distinct from 'soldier' in terms of meaning. Whereas under the *ancien régime*, a distinction had become established between 'soldier' (= 'mercenary') and the civilian world, in the 'era of revolutions', the dichotomy shifted to one between 'mercenary' (= 'foreigner in the service of princes') and 'defender of the fatherland' as conscript. According to the nationalists (in Germany, referred to as the Liberals), only the latter now could claim actually to be a true 'soldier'.[66] For this to happen, the 'fatherland' had to change into the 'nation'.

Meanwhile the process of further professionalisation continued. It was not until the nineteenth century that routine military service and training became common also in peace time throughout the year; only then did administrative control by the state offer the possibility to implement 'universal' conscription in reality – even if this was not practised as 'universally' as the term suggests. Only then was the military sphere in Prussia separated from the civilian sphere through the construction of barracks and the development of large exercise areas. Only then were the military/economic gaps in the state's all-pervading influence closed. These gaps had also opened up opportunities in the eighteenth century for military entrepreneurs to raise volunteer corps at their own expense and for their own benefit. But now the regular armies offered no place for profit-oriented old-style 'partisans'. This term no longer applied from the unit's commander in the context of the regular army. It now applied to the irregular fighter in the people's war. The term 'people's war' remained so broad, however, that any army practising conscription could describe itself as its agent.[67]

The people's war did not start with Napoleon. But, as the focus of the 'popular movements' directed against him, Napoleon marked the beginning of a redefinition of political semantics. In just the same way as the Marseillaise is a reminder of the people's mobilisation in defence of the fatherland, the Iron Cross owes its roots to plans for fighting a *guerrilla*-style people's war.

Notes

1. Nipperdey, *Deutsche Geschichte 1800–1866*, 11.
2. Schulze, *Staat und Nation*, 150–72, 202–3; Wehler, *Deutsche Gesellschafts-geschichte*, 1: 506–30; Koselleck, 'Einleitung', XV.
3. Rink, 'Preußisch-deutsche Konzeptionen'. See also Heuser, 'Small Wars in the Age of Clausewitz'.
4. Jessen, *Preußens Napoleon?*, 203–59.

5. Hahlweg, *Preußische Reformzeit*, 7–8.
6. Kunisch, *Der kleine Krieg*, 7.
7. Rink, 'The Partisan's Metamorphosis'; *Vom Partheygänger*; 'Der kleine Krieg', 356–7.
8. Luh, *Der Große*.
9. Grandmaison, *La Petite Guerre*, 2.
10. Satterfield, *Princes, Posts and Partisans*; Tóth, 'Régularité et irrégularité'; Picaud-Monnerat, 'La guerre de partis'; *La Petite guerre*, 122–30.
11. Picaud-Monnerat, 'Partisan Warfare, War in Detachments', 317–24; Rink, 'Der kleine Krieg', 383–4.
12. Scharnhorst, *Handbuch für Offiziere*; *Militairisches Taschenbuch*.
13. Höhn, *Revolution–Heer–Kriegsbild*, 233–366.
14. Emmerich, *Der Partheygänger*; Ewald, *Dienst der leichten Truppen*; Valentini, *Abhandlung*; La Roche Aymon, *Ueber den Dienst der leichten Truppen*; Schels, *Leichte Truppen*; Decker, *Der kleine Krieg*, 1828.
15. Voß, *Der kleine Krieg*, iii.
16. Rink. *Partheygänger*, 241–5.
17. Rink. 'The Partisan's Metamorphosis', 30; Coutau-Bégarie, 'Guerres irrégulières', 47.
18. Schiller, Friedrich. *Die Räuber*, Act 1, Scene 2. http://gutenberg.spiegel.de/buch/3339/1.
19. Möller, *Fürstenstaat*, 452–66; Echternkamp, *Der Aufstieg*, 149–59.
20. See the contribution of Beatrice Heuser below.
21. Sikora, 'Scharnhorst'.
22. Rink, 'Graf Wilhelm von Schaumburg-Lippe'.
23. Scharnhorst, *Taschenbuch*; cf. Frederick II's own instructions, *Des Königs von Preussen*.
24. Rink, *Partheygänger*, 221–41; Jessen, *Preußens Napoleon?*, 11f.
25. Berenhorst, *Betrachtungen*, 2nd ed. 2: 326–9; Jessen, *Preußens Napoleon?*, 216–18.
26. On the tirailleurs' enthusiasm as a factor of combat effectiveness according to the spirit of the national-socialist era: Höhn. *Revolution*, 244–5. Very critical: Mönch, 'Rokokostrategen', 87–95.
27. Scheipers, 'Status and Protection of Prisoners of War'.
28. Planert, *Mythos vom Befreiungskrieg*, 386–404.
29. Hahlweg. *Preußische Reformzeit*, 58–60.
30. Forrest, 'The Ubiquitous Brigand'; Boycott-Brown, 'Guerrilla Warfare'.
31. Planert. *Mythos vom Befreiungskrieg*, 424–37; Heitzer, *Insurrectionen*,122–7.
32. [Cölln]. 'Briefe', 94.
33. Rink, 'Ein Patriot und Partisan'; Bock, *Schills Rebellenzug*, 46.
34. On the German reporting of the Peninsular War, see Wohlfeil, *Spanien*, 19–39, 102–63; Esdaile, *The Peninsular War*, 37–86, 295–308.
35. Broers, *Napoleon's Other War*; Reynaud, *Contre-guerilla*; Lepetit, 'Soumettre les arrières de l'armée'.
36. Wohlfeil. *Spanien*, 102–63.
37. Rößler, *Graf Stadion*, 1: 293.
38. Botzenhart and Hubatsch, *Freiherr vom Stein*, 2: 808–12; Ibbeken, *Preußen 1807–1813*, 34–5; Duchhardt, *Stein*, 224–5.
39. Kunisch et al., *Scharnhorst*, 5: 434–42, 665–78.
40. Spanish manifestos in Wohlfeil, *Spanien*, 299–317. Gneisenau's 1808 memorandum in Thimme, 'Zu den Erhebungsplänen'; Clausewitz, 'Bekenntnisschrift von 1812'.
41. Schmidt, *Erinnerungen*, 1: 123.
42. Stamm-Kuhlmann, *König in Preußens großer Zeit*, 308, 332.

43. Veltzke, 'Zwischen König und Vaterland'.
44. Bärsch, *Ferdinand v. Schill's Zug*, 1–38, 218–20 (FN 38).
45. Binder von Krieglstein, *Ferdinand von Schill*, 147–55; Bärsch, *Ferdinand v. Schill's Zug*, 56.
46. Bärsch, *Ferdinand v. Schill's Zug*, 73–115; Binder von Krieglstein, *Ferdinand von Schill*, 160–99.
47. Zimmermann, *Der Schwarze Herzog*.
48. Memorandum of 8 August 1811 printed in Pertz, *Leben des Gneisenau*, 2: 106–42.
49. Ibbeken, *Preußen*, 108–11; Stamm-Kuhlmann, *Friedrich Wilhelm III*, 274.
50. Esdaile, *Fighting Napoleon*, 82–8, 130–59; Fraser, *Napoleon's Cursed War*, 338–9; Martínez Laínez, *Como lobos hambrientos*, 83–95.
51. Clark, *Preußen*, 400–5.
52. Aschmann, *Preußens Ruhm*, 230–60.
53. 'An Mein Volk' and 'An Mein Kriegsheer', *Schlesische privilegierte Zeitung*, 20 March 1813. Drafts in: Geheimes Staatsarchiv Preußischer Kulturbesitz, Rep 92, N 45, No. 2, fol. 2–4. Discussed in Rink, *Partheygänger*, 322–3; Aschmann, *Preußens Ruhm*, 276–80.
54. Bald, 'Wehrpflicht'.
55. Decrees printed in Frauenholz, *Entwicklungsgeschichte*, 5: 141–57.
56. Decrees in Frauenholz, *Entwicklungsgeschichte*, 161–75. Further: Rink, *Partheygänger*, 323–36; Langendorf, 'Landwehr et Landsturm'.
57. Friederich, *Die Befreiungskriege*, 1: 122f.; Ibbeken, *Preußen*, 398f.
58. Nitschke, *Die Preußischen Militärreformen*, 184–6.
59. Rink, *Partheygänger*, 329f.
60. 60 Pertz, *Leben des Gneisenau*, 2: 106–42.
61. Aschmann, *Preußens Ruhm*, 280–3.
62. Arndt, *Katechismus*, 14.
63. Körner, 'Männer und Buben'. On the 'male' coding of conscription, see Hagemann, *Männlicher Muth*, 271–350.
64. Walter, *Preußische Heeresreformen*, 235–469.
65. Heuser. 'Small Wars in the Age of Clausewitz'
66. Rotteck, *Ueber Stehende Heere und Nationalmiliz*, IIIf.; *Conversations-Lexikon*, 9: 208–15.
67. Rink, 'The Partisan's Metamorphosis'; "Preußisch-deutsche Konzeptionen".

Poachers turned gamekeepers: A study of the guerrilla phenomenon in Spain, 1808–1840

Mark Lawrence

Department of History, Newcastle University, Newcastle-upon-Tyne, UK

This article modifies the associations made by historians and political scientists of Spanish guerrilla warfare with revolutionary insurgency. First, it explains how the guerrilla phenomenon moved from a Leftist to a reactionary symbol. Second, it compares the insurgency and counter-insurgency features of the Carlist War (1833–1840) with those of the better-known Peninsular War (1808–1814). Third, it shows how erstwhile guerrilla leaders during the Carlist War made their expertise available to the counter-insurgency, in a socio-economic as well as military setting. This article revises the social banditry paradigm in nineteenth-century Spain in the under-researched context of Europe bloodiest nineteenth-century civil war.

Guerrillas in Spain from the Peninsular to the Carlist War

The Spanish guerrilla gave the Peninsular War its most enduring image, and it is an image which has equated the Spanish 'people' with the guerrillas fighting, and defeating, the occupying French army. The Peninsular War gave the English language the term 'guerrilla', meaning 'small war', but it certainly did not invent the phenomenon. The Dutch used irregular resistance against the Spanish Empire in the Eighty Years War, while even peninsular Spain itself saw significant episodes of popular resistance during the War of Succession (1701–1714) and during the War of the Pyrenees (1793–1795).[1] But the Peninsular War was certainly on a different scale: the 'small war' became truly 'big'. No part of French-occupied Spain outside of the major urban centres was far from some degree of irregular resistance, while some areas of the country, especially Navarra, were far more under the sway of the guerrillas than under the French. It was also big because the guerrilla phenomenon reverberated across far more dubious militancies linked to pillage, murder, and banditry.

During the nineteenth century, Spanish historians understandably overlooked these unwelcome features and instead celebrated the morality tale of their nation having risen against Napoleon's perfidious designs on Spain. The guerrillas literally embodied national sovereignty at a time when Spain's rightful king was a prisoner of the French. This morality tale united Spanish writers of both Leftist and Rightist sympathies, as the Leftist Liberal message of national sovereignty was as plausible as the Rightist message of popular defence of Throne and Altar.[2] And it was all the more satisfying as the tale ended in victory (albeit one which could not have been delivered without the Spanish regular army and its Anglo-Portuguese allies).

During the twentieth century, in response to the innovative historical techniques first of the Annales School and then of Cold War Marxist social history, historians turned the nationalistic image into a political one. A leading progressive historian was Miguel Artola, at the vanguard of the resilient 'bourgeois revolution' model, which modernised the 'people's war' discourse handed down since the nineteenth century along Marxist socio-economic lines.[3] For Artola, popular support for the war against the French was as an essential precondition which defined the 'revolutionary' and 'total' war against Napoleon.[4] Anglophone readers may find a similar, but more nuanced argument from the late Ronald Fraser, for whom the 'guerrilla was the crystallization of the villagers' will to resist. It was the only form of combat available to the weak against the strong. The guerrilleros were the villagers' kith-and-kin.'[5] The fact that so much of the guerrillas' militancy amounted to interpersonal violence and expropriation was explained either as the vicissitudes of irregular warfare, or systematised as 'social banditry'. Thus, according to the late Eric Hobsbawm, the violence of the impoverished rural population against a growing capitalist class was a typical feature of partially modernised but generally backward societies like Spain.[6]

Hobsbawm's thesis has, of course, been extensively revised.[7] Yet even though the image of the bandit is ambiguous at best, its interconnectedness with irregular resistance makes early nineteenth-century Spain one of its most prolific case studies. And despite the progressive representation of irregular resistance in the Peninsular War, the same phenomenon became an agent of reaction as Spain underwent the last great conflict of the pre-industrial age, the First Carlist War of 1833–1840. This conflict saw the centralising state and its capitalism triumph against the economically marginalised rural populations of the regions who formed so much of the guerrilla forces of the reactionary Carlists.[8]

Let us first provide a summary of the two conflicts. The Peninsular War began when Spanish 'Patriots' broke the increasingly fraught alliance with France when Napoleon subjugated the Spanish royal family. Initially poorly organised Patriot forces turned on the French occupiers, and despite the lowest point in their fortunes being reached in early 1812, Spanish regular and irregular forces thereafter came under effective British control in a joint and successful Anglo-Portuguese-Spanish liberation of the Peninsula. The First Carlist War began a good two decades later when the erstwhile prisoner of Napoleon, King Ferdinand VII, died, triggering a dispute within the Bourbon dynasty about whether the reactionary faction under Prince Charles (Don Carlos, hence the word 'Carlist') or the liberal faction under the infant princess, Isabella and Ferdinand's widow,

María Cristina (hence the word 'Cristino'), should reign. In the seven-year civil war that ensued, the largely regular Cristinos eventually vanquished the largely irregular Carlist forces, although not before liberal politics had been radicalised by the fighting and perhaps some 4% of Spain's population had been killed.

The evolution of Spain from a national (Peninsular) to a civil (Carlist) war is thus problematic from political science and progressive point of view. How could the guerrilla phenomenon move from a movement of national liberation[9] to one of reaction against nation-building? This question is all the more pressing because the Carlist War had an even stronger irregular and counter-insurgency dimension than the Peninsular War. The fact that rural guerrillas rose as reactionaries during 1820–1823 and again after 1833 was problematic for Spain's early liberals who claimed to represent the Spanish people. This was a far cry from the claims made by Spain's first liberalism during the Peninsular War. Then, one of its leading figures, the young aristocrat, Count Toreno, blamed reactionaries (called *serviles*) for the defeats sustained in the field by Spain's regular armies, but said that the people (viz. the guerrillas) were redeeming the nation, and sympathetic foreign observers shared his view.[10] In the more strident terms of a Spanish Jacobin, as Spain had been abandoned by her grandees and generals, it was left to the 'nation' to resist the monster Napoleon ('Spain today is what revolutionary France used to be').[11] Another Liberal, anguished by the flagging war effort in his native Galicia, exhorted villagers to rejoin the fight, for it was the Spanish people who had 'declared war in the first place'.[12] Such was the power of the popular resistance trope that even regular officers played to its strengths. General Ballesteros became a hero of the Left for defying the Duke of Wellington's Supreme Command of Allied forces in the Peninsula.[13] Of course, all these attempts at politicising Spain's popular resistance to Napoleon spoke far more of the liberals' political ideology than of reality. The liberals, resentful of the eighteenth-century militarisation of Spain's Bourbon monarchy, and resolute in their belief in the rights of the individual against arbitrary power, cast a suspicious eye towards Spain's Bourbon military, even though the rump of this had sided with the Spanish Patriots in 1808 and subsequently took to the field against appalling odds, never abandoning the struggle against Napoleon.[14] Thus the guerrilla phenomenon originally offered a ready-made counterweight to absolutism; by contrast, during 1820–1823 and again during 1833–1840, it was a counterweight to modernising liberalism. The irony of this retrograde evolution lies not just in the fact that the classic home of the modern 'small war' offers a glaringly Rightist narrative in a field of research dominated by Leftist social historians and political scientists, but also in the fact that Spain's liberals themselves set great store by claiming to represent 'the people'.

Even though the Peninsular War ended badly for Spain's first liberal movement, several guerrilla leaders initially continued to identify with the cause of the Left. The central plank of liberal reform was the advanced Constitution of 1812, which the Spanish king annulled in 1814 once he was released from French captivity at the end of the war. Several prominent guerrilla veterans had been fellow travellers of the liberal movement and now saw their careers suffer accordingly. The most successful guerrilla leader, Francisco Espoz y Mina, dubbed the 'king of Navarra' by his French enemies, expected to be rewarded with the vice-royalty of that autonomous kingdom in Spain. In fact, the king repulsed his entreaties, and Espoz raised a rebellion in the name of the Constitution of 1812.[15]

Meanwhile, other guerrilla leaders tainted by liberalism followed Espoz in his forlorn bid to force the Constitution on the king and rescue their careers, including Juan Díaz Porlier in 1815, and Luis Roberto Lacy and Vidal, both in 1819. When the king's resistance to liberalism was finally overcome, it was due to the raising of a regular army (led by Colonels Riego and Quiroga over the winter of 1819–1820). The regime which this liberal rebellion imposed on Spain was called the Triennium, lasting from 1820 until 1823. Almost immediately the triumphant liberal regime was met by resistance by reactionaries in the Spanish countryside (known as the Royalist War of 1820–1823). The reactionaries were particularly strong in regions of Spain that resisted the liberals' programme of advancing rural capitalism and subordinating the traditionally powerful Catholic Church to state power. These included the rural Basque country, especially Navarra (where strong local autonomy and popular religiosity were matched by a satisfied pre-capitalist peasant class), the interior of Catalonia, Aragón, and Valencia, and other, patchier, areas across Spain besides. Most of this reactionary violence was a template for the far more serious civil war, which erupted in 1833. Both the previous Royalist War of 1820–1823 and, especially, the First Carlist War 1833–1840 had strongly irregular and counter-insurgency dimensions. And instead of irregular 'national liberators' fighting against regular armies, these later wars saw the roles reversed. Now revolutionary armies and, especially, the liberals' own paramilitary citizens' force, the National Militia, fought against irregular forces opposed to liberal programme of nation-building. This role reversal is all the more striking as it was the major difference between two conflicts which otherwise shared many similarities.

Insurgency and counter-insurgency similarities between the Peninsular and Carlist Wars

The first similarity is human, and one that is significantly elitist in nature. A majority of Peninsular War guerrilla veterans who either retained, or returned to, elite military and political commands in 1833, and who were not intimately associated with the three Carlist heartlands (see below), fought on the Cristino rather than the Carlist side. Leftist historians have celebrated this Left-political recovery of guerrilla leadership as a welcome corrective to the Royalist War and a partial return to the national liberation promise of the Peninsular War.[16] And certainly, even two prominent Peninsular War veterans who *were* intimately associated with the Basque country (Espoz y Mina, Jaúregui), fought on the Cristino side.

This elitist observation, however, obscures a far more profound subaltern phenomenon, which was reactionary rather than liberal in nature. The vast majority of irregular combatants who took to arms from 1833 did so for Carlism rather than liberalism. Indeed, this phenomenon was not restricted to the insurgent heartlands, but also applied by degrees to the patchy areas of Carlist support. In many cases, subalterns joined Carlist guerrilla leaders who boasted of their exploits against the French a generation earlier.

No exact figures exist but they range from commanders of large-scale (more than 3000 effectives) to medium (500–3000) and small-scale (fewer than 500) forces. The most prominent veteran commanding large-scale forces was Jerónimo Merino (alias El Cura), who from his traditional support base in rural Burgos province during the early stages of the war carried out a failed march on Madrid.[17] In rural Castellón (the eastern zone), a veteran commanding medium-scale forces was José Miralles (alias El Serrador), an illiterate woodsman who besieged Cristino militia and army garrisons.[18] In rural New Castile (now Castilla la Mancha) a veteran commanding small-scale forces was Manuel Adame (alias El Locho), of even humbler social origins than Miralles.[19] All three were also veterans of the Royalist War. The most famous Carlist guerrilla commander of the Carlist War, Tomás de Zumalacárregui, commander-in-chief of Carlist forces in the Basque country between 1834 and 1835, was also a veteran of both wars, but at a subordinate rank during 1808–1814. Ramón Cabrera, commander-in-chief of Carlist forces in the Maestrazgo between 1835 and 1840, was younger and new to fighting in 1833. While the Carlist order of battle varied across Spain, three common categories may be identified in its officers. The first comprised the veterans of the Royalist War (1820–1823), the second, most reactionary, category comprised those who rose to command during the 1833 rising, while the third category comprised foreign officers. The last included both long-standing members of the Spanish army hierarchy with careers stretching back to before 1833 (to as early as the 1790s in the case of Penne de Villemur) and international volunteers who came to Carlist Spain after 1833. Many of these – especially the sizeable German contingent – brought technical application to the Carlist war effort.[20]

The second similarity is geographical, both at regional and local levels. Regionally, the only area of Spain where genuinely popular resistance to Napoleonic occupation took place, rural Navarra, was also the first heartland of popular Carlist resistance after 1833. This was no coincidence, but rather the product of identifiable socio-economic conditions threatened by first Napoleonic and later Cristino–liberal reformism, coupled with a popular psychology of armed resistance and organisation handed down through generations. At the summit of these similarities was one significant human difference. Espoz y Mina, Patriot commander of the Peninsular War Navarra and failed pretender for viceroy at the war's end, finally became viceroy in 1834, but ironically as a Cristino liberal fighting *against* his veterans in Navarra, and the sons of his veterans, who flocked to the reactionary Carlist movement in 1833.

Spain's Leftists claimed Espoz as a nation-builder, a 'man of his fatherland (*patria*)', in contrast to Zumalacárregui who was merely a 'man of his mountains'.[21] But Espoz's opponent, Zumalacárregui, defeated a succession of Cristino viceroys, including the erstwhile 'king of Navarra'. The first stage of Zumalacárregui's strategy began between the winter of 1833 and the following summer with the employment of his peasant soldiers in a guerrilla war of surprise and ambush. During this first stage his army lacked the strength to threaten urban centres, particularly the prized Vizcayan capital, Bilbao. Zumalacárregui's second strategic stage ran from the summer of 1834 until the winter, and involved controlling rural territory and communications. The ensuing third stage involved taking Cristino forts and fortified towns, which culminated in the first Carlist siege of Bilbao during the spring and summer of 1835, during which Zumalacárregui was fatally wounded.[22]

There were also regional nuances between the Peninsular and Carlist Wars. The second heartland of Carlism, the Maestrazgo and its adjoining Aragonese and Valencian uplands, was certainly a scenario for Patriot irregular warfare during the Peninsular War (although vigorously pacified by French blockhouses). Yet, it was of a far greater magnitude during the Carlist War. The Carlist effort in the Maestrazgo was led from 1835 by Ramón Cabrera and went through three incremental stages. During the first stage, 1833–1835, the Carlist effort here was dispersed by the Cristinos and confined to rural guerrilla warfare. The second phase, 1836–1837, saw the breakthrough as Cabrera expanded the rural boundaries of the Carlist control and turned it into a military state. The third phase, 1838–1839, in contrast to the stagnation in the Carlist Basque country, saw the Maestrazgo reach its apogee in terms of geography, military victories, and recruitment. The August 1839 Peace Treaty in the north orphaned these achievements and sent Cabrera into exile the following year, ending the First Carlist War.[23]

A further nuance was in the third Carlist heartland, rural Catalonia, which intermittently expanded to link with the Maestrazgo. Rural Catalonia, like the Basque country, also had a strong tradition of popular armed mobilisation in the form of the *somatenes*, or local militia (in Navarra their equivalent were called *requetés*). But the Carlist zone in rural Catalonia was more hemmed in by terrain and a particularly vigorous Cristino counter-insurgency during 1833–1834. It also had to wait longer for effective Carlist leadership to emerge (under first Guergué and then the Conde de España).[24] Beyond these regions, the geographical distribution of Carlism was patchy and changing, although support for Carlism was generally strong in northern Castile, Asturias, and Galicia. Rural pockets were often held by what the Cristinos called 'enemy agents' (*facciosos*), but many of these 'Carlists' were really bandits looking for political cover, or deserters from the Cristino army, which will bring us to discuss another similarity between the Peninsular and Carlist Wars (see below).

In terms of geographical similarities at local level, both Peninsular War Napoleonic occupation forces and Carlist War Cristino garrisons retained control of large urban centres. The smaller and remoter the urban centres, the more their control was likely to be contested by Patriot and Carlist irregulars. The largest urban centre to be captured by Carlists during the war was Córdoba in 1836, and the impetus for this conquest came not from local Carlists, who in Andalucía were comparatively few, but from an invasion launched from the Basque country. Even in the Basque country, the strongest of the three heartlands, Carlist control was hemmed into small market towns and proto-industrial centres. Even here Cristino counter-insurgency operations forced the frequent relocation of the Carlist 'capital'. Between the Carlist Pretender's arrival in Spain in July 1834 and September 1836, the Carlist seat of power was forced to relocate no fewer than nine times, between Estella, Iturmendi, Oñate, Durango, Elorrio, Villarreal de Guipúzcoa, Villafranca, Goyaz to Azpeitia, and Tolosa.[25] Moreover, with the exception of an unruly Carlist occupation of Bilbao in the first few weeks of the war, all the provincial capitals in the Basque country (Bilbao, San Sebastián, Vitoria, Pamplona) remained in Cristino hands throughout.

The rural *versus* urban divide of the Carlist War was well noted by contemporaries. Charles Henningsen, an Anglo-American Carlist volunteer thought Spanish civil wars were essentially a conflict between liberal towns and royalist countryside.[26] Even in areas of Spain where Carlism was weak, such as Málaga, the rural population frequently identified with Carlism in reaction to unpopular liberal economic reforms, which they associated with the towns. In an incident in May 1834, a militiaman sent out from Málaga was ambushed by villagers shouting 'this one's a liberal like all the others in his unit: die'.[27] The urban–rural divide can be overstated, however. Even in urban settings artisans threatened by recurrent liberal abolition of the guilds, and redundant officeholders (*cesantes*) could be drawn to Carlism, while, by contrast, villagers who had benefitted from the liberal property revolution would correspondingly support constitutionalism.[28]

The Carlist insurgency

Zumalacárregui's strategy mirrored that of Espoz on the same territory during the Peninsular War, namely in the gradual militarisation of his guerrilla volunteers, and in the suffering this imposed on non-combatant populations. Within weeks of the first rising in October 1833, Zumalacárregui made himself Carlist commander-in-chief of Navarra, at the head of a growing army of local peasants using guerrilla tactics, which turned Cristino garrisons and counter-insurgency manoeuvres into a pale imitation of those of the French a generation earlier.[29] These guerrilla tactics were overborne throughout 1834 by Zumalacárregui's painstaking formation of cavalry and artillery units, eroding Cristino superiority in these arms.[30] By the time of Zumalacárregui's death in June 1835, a Carlist Royal Army had been formed in the Basque country, fluctuating around 30,000 men, which was sustained by conscription (despite the Carlists' persistent terms of address to 'volunteers') and growing standardisation of uniform (including the famous 'boina', or beret) and supplies (including the militarisation of working conditions in rural arms factories).[31] These measures, which could not have been substituted for guerrilla warfare on its own, won the Carlists domination of most parts of the Basque countryside outside of provincial capitals. Hence this outcome presents us with a similarity and a paradox in comparison to the Peninsular War. The similarity is that Espoz y Mina in his war against Napoleon also militarised his effectives in a similar manner. In both cases, guerrillas evolved into formally militarised armies – Mao would have approved. Both insurgencies were militarily at their most effective after this transition. Regular warfare, it would seem, was more effective than irregular 'people's war' (indeed, the Carlists would finally be brought to peace terms in 1839 because of the failure of their guerrilla efforts beyond the Basque country).

In both case studies this process of militarisation impacted negatively on civilian communities which faced not only conscription, but also financial and physical restrictions on their freedom which belied the liberation propaganda of the insurgents. In 1813, a Zaragoza diarist reported Espoz's troops to be treating his city 'as if they were at war with the population', inflicting arbitrary injustice and violence on civilians with apparent callousness.[32] Equally, when the Carlist Gómez Expedition occupied Córdoba in October 1836, the streets descended into violent lawlessness and local liberals were fined or expropriated.[33] Espoz behaved so in the name of liberty, whereas the Carlists did so in the name of religion, yet the effect on non-combatants was much the same. Interpersonal violence between Spaniards characterised the Peninsular War, but was cloaked by the 'national' nature of the struggle.[34] The liberals in the 1830s, too, called themselves patriots, but now the enemy were fellow Spaniards, which meant that interpersonal violence has dominated histories of the Carlist War to a far greater extent than is palatable in histories of the 'national' struggle against Napoleon. Liberal histories of Carlist violence, in particular, have tended to personalise the phenomenon, especially with regard to the ruthless commander-in-chief of Carlism's eastern heartland after 1835, Ramón Cabrera, thereby conveniently overlooking the progressive economic reforms which provoked it.[35] In like manner, nationalistic histories of the Peninsular War (until Miguel Artola's work) conveniently overlooked the enlightened reforms of the French occupiers, preferring instead to focus on Patriot initiatives and to personalise the enemy (most famously, in the depiction of the 'intruder king', Joseph I, as a hapless drunkard, 'Pepe Botellas').[36]

To illustrate the dimension of interpersonal violence inherent in the Carlist Wars' guerrilla phenomenon, let us analyse not one of the heartlands but two patchy areas of Carlist support, Old and New Castile in mid war (1836), as opportunities for irregular warfare beyond the nominal front lines were far greater. At least 24 medium- and large-scale bands were active in this area. Four main groups can be identified in these guerrilla bands. Firstly, they contained deserters fleeing the appalling conditions of the Cristino army, which by now had been virtually paralysed by revolution. Secondly, they contained conscripts waiting to join the Cristino army (*mozos*), but whom the Carlist bands had collected from villages and small towns they invaded, either forcing the men into their ranks, or bribing them with promises of licence and pillage. Thirdly, several were outright criminals who found campaigning to be an excellent motor for crime. And fourthly, these bands contained agricultural labourers who had fallen on hard times due to war-ravaged harvests, or had been seen their customary rights eroded by the new capitalist reforms which was concentrating landownership in fewer hands. A final category, whom we might term 'adventurers', were simply persuaded to take to the hills to escape a life of poverty and drudgery.[37] Thus, with the exception of the fourth category, contingent rather than socio-economic or ideological motives drove villagers to join Carlist bands. That said, ideologically-committed Carlists sometimes turned these contingent motives to their advantage. Andalucía, for example, where Carlism was weakest, nonetheless saw attempts by Carlist emissaries to turn the bandits, smugglers, and deserters in the region's lawless Serranía de Ronda into a guerrilla force.[38] But the contingent motivations do suggest a major similarity with the Peninsular War, where even Leftist works have conceded a 'grey line' between militancy and criminality.[39] But whereas irregulars resisting state power during the Peninsular War could reinvent themselves as Patriots, their successors during 1830s did so as Carlists.

Cristino counter-insurgency

After their initial counter-insurgency campaigns in the Basque country failed to nip the Carlist rising in the bud over the winter of 1833–1834, the Cristinos switched to a counter-insurgency doctrine of blockhouses, or fortified lines, which had been successfully pioneered by the French during the Peninsular War.[40] But unlike the French, or at least unlike the French until 1812, the Cristinos had too few troops to make this scheme effective, and by the time enough troops were available, their efficacy was undermined by political radicalisation (the Spanish army had been purged and kept deliberately small in 1833 in order to neutralise any coup attempt by Carlists; by 1837 the Cristino army had grown to over 200,000 from a pre-war strength of 40,000, but by then radical liberal revolutions in the urban centres were derailing Cristino military strategy from within).[41] The Cristinos' extended lines of communication meant that, for example, 10 liberal soldiers were needed to transport one battle casualty to safety. Soon after being appointed commander-in-chief of the Cristino Army of the North in 1835, Luis Fernández de Córdova claimed that 19 out of every 20 messengers he sent out gave up their intelligence to the Carlists, the brave twentieth man perishing instead. This was an exaggeration, but one born of despair.[42] Increasingly it was the liberals' own paramilitary force, the Urban Militia (expanded and renamed National Guard in 1835 and National Militia in 1836), which could provide the only effective defence of Cristino urban centres cut off in Carlist heartlands. And even these were often so intimidated by rural guerrillas as to be forced under Cristino army command on repeated occasions. Espoz y Mina, as captain-general of Catalonia in July 1836, initiated a desperate three-pronged counter-insurgency strategy. Local militia forces were ordered to fight Carlist invaders whenever the latter's numbers did not exceed one-half of local available defence forces. Crippling fines were to be imposed on local authorities that failed in this order. Even a deforestation programme was ordered in order to deprive Carlist guerrillas of natural cover.[43]

Another similarity with the Peninsular War lay in the question of quarter. During 1809 the Patriot provisional government (Junta Central) went to great lengths to regularise the status of Patriot irregulars in the hope that the French would recognise those they captured as lawful combatants entitled to quarter. From 1833 the insurgent Carlists did much the same, and in April 1835 an internationally brokered treaty gave quarter to regular combatants surrendering in the Basque country. The fighting here had largely evolved from irregular to regular warfare by then, and Carlist forces outside of the Basque country falling into the hands of Cristino counter-insurgency forces could still be lawfully executed (and even in the post-treaty Basque country captured prisoners were often still executed in a spiral of reprisals). In reality, the sheer numbers (sometimes all the menfolk of a village) of Carlist irregulars forced Cristino counter-insurgency forces to issue amnesties, although this as a consequence led to the capture of the same irregulars more than once. The practice of Carlist irregulars to 'change their shirt' began in the Basque country. 'Shirt-changing' described the practice of Carlist insurgents losing, deserting, or retreating in the face of superior Cristino forces and returning to their homesteads where they could conceal their weapons and choose their moment to return to fight from a position of advantage. Shirt-changing was at its most common in the popular Carlism of the Basque country, particularly during the first 18 months of the conflict while the Carlist forces were being regularised. But it also happened in the patchy areas of Carlism beyond the heartlands, and Cristino officers assumed that these guerrillas were either being forced to fight by local elites sympathetic to the reactionary cause, especially priests, or were in league with organised crime.[44]

Erstwhile Cristino insurgents were now at the vanguard of brutal counter-insurgency. Espoz y Mina had been the hammer of royalist guerrillas in Catalonia in the early 1820s. When Espoz in October 1835, having failed to stem Zumalacárregui's insurgency, was moved to the captaincy-general of Barcelona, he used similar tactics against Catalan Carlists. He showed no understanding of the 'social banditry' behind Catalan Carlism and instead tarred his enemy as fanatical brigands bereft of civic education.[45] Associations of Carlism with banditry and priests are long-standing, and they lent the Cristino side moral legitimacy. The great liberal poet and pessimist, Mariano José de Larra, in his *Nadie pase sin hablar al portero* at the start of the war dubbed the northern Carlists bandit priests.[46]

The Cristinos, in waging their nation-building civil war, thus shared the language as well as the tactics of their national enemy a generation earlier. Espoz's draconian counter-insurgency measures matched anything the French did. On 29 November 1835, he imposed a state of siege in Catalonia, with measures including collective punishment on communities invaded by Carlists, ranging from summary executions of heads of households harbouring and helping enemy agents to the confiscation of property.[47] Nor was Espoz the only insurgent veteran who now turned to Cristino counter-insurgency. Gaspar de Jaúregui (alias 'El Pastor'), guerrilla veteran of the Peninsular and Royalist Wars, was in 1833 a liberal exile, just like Espoz, but now he was recalled to his native Guipúzcoa to lead a local Basque-speaking Cristino free corps (called 'chapelgorris') in counter-insurgency operations.[48] These too were remembered for their brutality against local insurgents, largely because they were themselves uniquely vulnerable to reprisals.[49] Another 'poacher turned gamekeeper' was Juan Palarea (alias 'El Médico') who was involved in a radical conspiracy at the centre of Cristino power in 1834, and would end the Carlist War as a regional governor in Andalucía charged with pacifying popular radicalism unleashed by the conflict.[50] He also led counter-insurgency operations against Carlists trying to break out of the Maestrazgo into New Castile, Palarea succeeding in an area where support for Carlism was patchy, unlike the Basque country.[51]

Cristino irregulars against Carlists

The few examples of Cristino forces waging irregular warfare are the exceptions that prove the rule. Some of the Carlist guerrilla bands in New Castile discussed above were countered by Cristino counter-guerrillas. Cristino Comandante General Jorge Flinter created a counter-guerrilla force of desperate men the enemy derided as *peseteros* ('money-whores' who earned one peseta per day). These were often armed with nothing more than shotguns, but they were nonetheless partially successful as they, unlike regular campaign troops, knew the local terrain and had their own intelligence networks.[52] As most of the Basque country from the summer of 1835 was under organised and contiguous Carlist control, the Cristinos expanded upon their blockhouse strategy by including an irregular dimension. As Cristino Spain was plunged into radical revolution, most markedly at the hands of the liberals' militia in September 1835, and then at the hands of subaltern army ranks the following summer, regular troops became at best unreliable for counter-insurgency operations. Rather, Cristino garrisons themselves were frequently on the defensive against Carlist invasions, most prominently the Gómez and Royal Expeditions of 1836 and 1837. Cristino operations against the Carlist stronghold of Navarra thus turned to guerrilla activity.

Martín Zurbano Baras (alias Martín Varea) was born in Logroño in 1788 into a peasant family of distant noble origin. He fought as a guerrillero in the

Peninsular War and in the liberals' National Militia cavalry during 1820–1823. In the Carlist War he commanded a force of liberal guerrillas called *contra-aduaneros* to which Carlist civilians in his path added the suffix 'of death', or *de la Muerte*, on account of the black flags these guerrillas placed on their lances. This small, irregular cavalry wrought havoc, sabotaging Carlist granaries and other supplies and engaging in hit-and-run attacks. In addition to their depleted supplies, the Carlists lost some 500 dead, wounded, or captured at Zurbano's hands during 1835.[53] The internationally brokered treaty giving quarter did not apply to Zurbano's forces as they were irregular, which meant that his men were less likely either to give or to be given quarter. As the fortunes of the Cristino regular army recovered under Espartero's leadership from 1837, Zurbano's Cristino guerrillas had an ancillary role to play in terrorising Basque civilians as part of Espartero's general offensive. In August 1838 Zurbano burnt Guevara to the ground after having failed to dislodge its Carlist garrison from the town's castle.[54]

Carlist civilians in the path of Zurbano's raids mounted round-the-clock watches against him, and several local Carlist efforts tried (and failed) to assassinate him. Zurbano claimed his military honour to be offended by civilian vigilantism, and decreed draconian beatings to these watchmen and public executions of their local mayors and priests. Yet Zurbano's irregular campaigning was counter-productive, as his raids served only to stiffen Carlist resolve. Carlist heads of family exempted from conscription into the Royal Army, who had already formed voluntary flying columns in response to the pillaging of Cristino troops, successfully used these as counter-guerrillas.[55]

The last significant Cristino guerrilla operation was even more ineffective, and has been subjected to ridicule in the Carlist literature.[56] José Antonio de Muñagorri, a native of Verástegui (Alava), began the war as the backbone of Carlist war-related industry, owning six factories in Guipúzcoa and Navarra. After his repeated petitions to the Carlist court for a compromise peace fell on deaf ears, he emigrated to the French side of the Pyrenees where he contacted Cristino agents and forged a plan to raise a guerrilla force of Spanish Basques. By November 1837 Muñagorri had become a salaried Cristino officer and on 18 April 1838 he raised his standard in Verástegui calling for *paz y fueros* ('peace and fueros'). The 'fueros' were the ancient Basque charters of self-government, a major stumbling-block to peace with the centralising liberals, and Muñagorri's gambit reflected the growing willingness in Carlist Spain in the wake of the failed Royal Expedition to sacrifice Don Carlos in return for Basque rights.[57]

From the outset, however, Muñagorri's guerrillas encountered difficulties with their new-found Cristino allies. Regular Cristino forces invading the Basque country resented how Muñagorri's guerrillas were attracting deserters from their units. General O'Donnell, writing on 26 October 1838, thus refused Muñagorri safe passage across his zone of command.[58] Financial backing and pressure from the Cristinos' principal allies, Britain and France, relaunched Muñagorri's guerrillas in November, this time in coordination with Jaúregui's free corps. But, again, regular Cristino units, including those led by the erstwhile insurgent, Jaúregui, showed nothing but derision for Muñagorri's 'army' and refused cooperation.[59] In the end, the peace treaty drawn up in August 1839 came as a result of Espartero's relentless regular campaign.

Conclusion

The insurgency and counter-insurgency features of both the Peninsular War and First Carlist War shared similarities in their origins and outcomes, and yet these have been represented differently. The trope of guerrillas is that of national liberators turned into guerrillas as wrongdoing reactionaries resisting nation-building, yet the subjective and contingent motivations of the guerrilla fighters remained similar. The cases of prominent guerrilla national liberators returning as nation-builders are the exceptions that prove the rule. Similarities also extend to the counter-insurgency. The First Carlist War applied and perfected the counter-insurgency model which the French – on account of stronger regular resistance – could not successfully carry out a generation earlier. The failure of isolated Cristino attempts at replicating Patriot guerrilla warfare underscored the relative strength of regular rather than irregular warfare in early nineteenth-century Spain. The counter-insurgency phenomenon matured during the First Carlist War and explained its most enduring innovation soon after the war's end, Spain's first centrally organised rural gendarmerie, the Civil Guard (*Guardia Civil*), much on the pattern, incidentally, of late seventeenth- and early eighteenth-century France, showing similar responses to similar problems.

Notes

1. Borkenau, *Spanish Cockpit*, 1–20; Roura i Aulinas, 'Jacobinos y jacobinismo.
2. I discuss this politicised historiography further in Lawrence, 'Peninsularity and Patriotism'.
3. Artola-Gallego, *Los orígenes de la España contemporánea*; *La burguesía revolucionaria*.
4. Artola-Gallego, 'La guerra de guerrillas'.
5. Fraser, *Napoleon's Cursed War*, 366.
6. Hobsbawm, *The Age of Revolution*, 146–8.
7. Sant Cassia, 'Banditry, Myth and Terror'.
8. Pérez Garzón, *Milicia nacional*, 10–89, 217, 260; Aróstegui et al., *Las guerras carlistas*, 150–1.
9. In fact, Peninsular War scholarship since the end of the Cold War has strongly revised the 'people's war' paradigm, above all thanks to the works of John Lawrence Tone and Charles Esdaile, but the persistence of the national liberation theme may be shown not just in the popular imagination but also in such general surveys of guerrilla warfare as Boot, *Invisible Armies*, 80–92.
10. Cepeda Gómez, *El Ejército Español*, 137–8; Diego García, 'Balance de un conflicto', 32.
11. Romero Alpuente, *El grito*, 19.

12. *Semanario Político*, No. 36, 843–50 in Saurin de la Iglesia, *Manuel Pardo*, 843–50.
13. Romero Alpuente, *Wellington en Cádiz*.
14. Esdaile, 'War and Politics in Spain'. The administrative apparatus of Bourbon Spain underwent militarisation during the eighteenth century, whereby offices of secretaries of state were increasingly filled by military men and the powers exercised by captains-general in provincial administration came to eclipse the respective authority theoretically held by government ministers. Cepeda Gómez, *El Ejército Español*, 144–5.
15. Iribarren, *Espoz y Mina*, 179–82.
16. Fraser, *Napoleon's Cursed War*, 420–1.
17. Artola-Gallego, *Memorias*, 2: 223–34; Melgar, *Pequeña historia*, 86–92; Pirala y Criado, *Guerra civil*, 1: 209–15.
18. Pirala, *Guerra civil*, 2: 327–9; Oyarzun, *Historia del carlismo*, 148.
19. Pirala, *Guerra civil*, 1: 341–6.
20. Santirso, *Joseph Tański*, 166; Córdova y Valcarcel, *Memorias*, 1: 242.
21. Risco, *Zumalacárregui*, 153.
22. Aróstegui, 'La aparición del carlismo', 106.
23. Remírez de Esparza, *El carlismo aragonés*, 32–6.
24. Pirala, *Guerra civil*, 2: 260–5; Santirso, *Joseph Tański*, 61.
25. Pirala, *Guerra civil*, 2: 69–71; Oyarzun. *Historia del carlismo*, 151.
26. Comellas García-Llera, *El trienio*, 312.
27. *Eco del Comercio*, 24 May 1834.
28. Artola, *Revolución burguesa*, 90–112.
29. Bellver Amaré, *Zumalacárregui*, 20–31, 51, 131–42, 181–201.
30. Córdova, *Memorias*, 1: 197.
31. Bellver, *Zumalacárregui*, 249; *Gaceta Oficial*, 5 July 1836.
32. Biblioteca Universitaria Zaragozana (B.U.Z.), Faustino Casamayor, Años políticos e históricos de las cosas más particulares ocurridas en la Imperial Augusta y siempre heróica Ciudad de Zaragoza, XXX, 1813: 27 September 1813 diary entry.
33. *Eco del Comercio*, 12 October 1836; Pirala, *Guerra civil*, 3: 242–7.
34. Esdaile, *Fighting Napoleon*.
35. Pirala, *Guerra civil*, 1: 249–51, 274–88; Cabello et al., *Guerra última*.
36. Lawrence, 'Peninsularity and Patriotism', 453–68; Artola-Gallego, *Los afrancesados*.
37. Pirala, *Guerra civil*, 3: 172–9.
38. Ibid., 1: 341–6; *Eco del Comercio*, 10 July 1835.
39. Fraser, *Napoleon's Cursed War*, 335–45.
40. Risco, *Zumalacárregui*, 74; Alexander, *Rod of Iron*.
41. Santirso, *Joseph Tański*, 100.
42. Córdova, *Memorias*, 2: 134–42.
43. Pirala, *Guerra civil*, 3: 46–9.
44. Coverdale, *Basque Phase*, 173–5, 294–308.
45. Artola-Gallego, *Espoz y Mina*, 2: 324.
46. *Revista Española*, 18 October 1833.
47. Pirala, *Guerra civil*, 2: 300–3; Artola-Gallego, *Espoz y Mina*, 2: 334–5.
48. Burgo, *La primera guerra carlista*, 41; Pirala, *Guerra civil*, 1: 298–309; Artola-Gallego, *Espoz y Mina*, 2: 224.
49. *Eco del Comercio*, 2 January 1836; Artola-Gallego, *Espoz y Mina*, 2: 296–7.
50. Córdova, *Memorias*, 1: 201–2; Pirala, *Guerra civil*, 2: 109–14; Lawrence, 'Las viudas de Comares'.
51. Pirala, *Guerra civil*, 2: 335–8; *Gaceta Oficial*, 5 April 1836.
52. Pirala, *Guerra civil*, 3: 172–9. The Carlist press often referred to all Cristino soldiers as 'peseteros', presumably in order to imply their moral and financial misery.

53. Ibid., 2: 256–60.
54. Ibid., 4: 623–5.
55. Ibid., 5: 391–2, 4: 32–44.
56. Oyarzun, *Carlismo*, 133–5.
57. *Eco del Comercio*, 16 October 1838; Pirala, *Guerra civil*, 4: 565–73.
58. Pirala, *Guerra civil*, 5: 182–6.
59. Ibid., 5: 186–91.

Small Wars in the Age of Clausewitz: The Watershed Between Partisan War and People's War

School of Politics & International Relations, University of Reading, UK

ABSTRACT Around the time of Clausewitz's writing, a new element was introduced into partisan warfare: ideology. Previously, under the *ancien régime*, partisans were what today we would call special forces, light infantry or cavalry, almost always mercenaries, carrying out special operations, while the main action in war took place between regular armies. Clausewitz lectured his students on such 'small wars'. In the American War of Independence and the resistance against Napoleon and his allies, operations carried out by such partisans merged with counter-revolutionary, nationalist insurgencies, but these Clausewitz analysed in a distinct category, 'people's war'. Small wars, people's war, etc. should thus not be thought of as monopoly of either the political Right or the Left.

After unwarranted claims have been refuted that Clausewitz had nothing to say about the lesser wars that have predominated in the world since 1945, much literature has in fact been produced on Clausewitz's writing on or relevance to 'Small Wars', 'Low Intensity Conflict', 'New Wars', 'Contemporary War', and even NATO's anti-terrorism strategy.[1] The main interest of Clausewitz's gallant defenders,

---

Translations are made by the author of this article unless the work's title is already quoted in English.

[1] Paul D. Hughes, *Small Wars, or Peace Enforcement According to Clausewitz* (Carlisle Barracks, PA: US Army War College Strategic Studies Institute 1996); Christopher Daase, 'Clausewitz and Small Wars', in Hew Strachan and Andreas Herberg-Rothe (eds.), *Clausewitz in the Twenty-First Century* (Oxford: OUP 2007), 182–195; Stuart

however, has been to demonstrate that what he wrote about war in general can be applied not only to classic inter-state war, but also to Small Wars, intra-state war, irregular war etc.[2] Some works about Clausewitz's work generally home in on his treatment of Small Wars, but usually in a more theoretical context, seeking theory-oriented lessons for the present and future.[3] There is one big exception, the article by the late Werner Hahlweg dating back to 1986, which still provides the best summary of what Clausewitz's *On War* had to say about People's War, more about which later.[4]

In this article, we shall reconstruct writing and thinking about Small Wars just before, during and just after the time when Clausewitz was active, to gain a better understanding of what both he and his contemporaries writing on such matters had in mind. This is worthwhile because any theoretical treatise, whether this be on war, or on politics, the society, economics or what have you, needs to be understood first of all in the historical context in which it was formulated. Only if we know which concrete historical examples a writer has in mind when he or she tries to establish general principles will we understand both their scope and their limitations, and the connotations and denotation of the terms used to describe them. We shall discover that Clausewitz stands at the watershed of two forms of 'Small War', even though both he and his contemporaries showed little explicit awareness of this sea-change.

We thus have to start with definitions of the term 'Small War', and, for reasons that will become apparent, of the terms 'Partisan Warfare', and 'People's War', that are contemporary to Clausewitz's writing, and with the historical instances that the Prussian was aware of. It will become apparent that by and by, there was a shift in meanings.

Kinross, 'Clausewitz and Low-Intensity Conflict', *Journal of Strategic Studies* 27/1 (March 2004), 35–58; Eva Strickmann: *Clausewitz im Zeitalter der 'neuen' Kriege* (Berlin: Galda 2008); Antulio Echevarria, *Clausewitz and Contemporary War* (Oxford: OUP 2007); Erasmus Beckmann, *Clausewitz, Terrorismus und die NATO Antiterrorstrategie*, AIPA 3/2008 *Arbeitspapiere zur Internationalen Politik und Aussenpolitik* (Cologne Univ. 2008).
[2]Christopher Bassford, 'In defence of Clausewitz', *Times Literary Supplement*, 15 Jan. 1993; Colin S. Gray, 'Clausewitz Rules, OK? The Future is the Past – with GPS', *Review of International Studies* 25 (1999), 161–82.
[3]Jon Tetsuro Sumida, *Decoding Clausewitz: A New Approach to On War* (Lawrence: UP of Kansas 2008); Herfried Münkler, *Über den Krieg: Stationen der Kriegsgeschichte im Spiegel ihrer theoretischen Reflexion* (Weilerswist: Velbrück Wissenschaft 2008), 75–90.
[4]Werner Hahlweg, 'Clausewitz and Guerrilla Warfare', in Michael I. Handel (ed.), *Clausewitz and Modern Strategy* (London: Frank Cass 1986), 127–33.

The First Manifestation of Small Wars: 'Partisan Warfare' as Special Operations

The late sixteenth and early seventeenth centuries, a period marked by a great crisis which led to prolonged and endemic wars throughout Europe,[5] saw a pronounced departure from what had previously been thought of as regular warfare, war ruled by widely recognised conventions, which were not yet as yet codified in any internationally agreed norms. Many wars of this crisis period stood out for their atrocities and the use of irregular warfare side by side with regular battles. The Dutch Revolt against Spanish rule in what the Dutch call the Eighty Years War (1568–1648), and the overlapping Thirty Years War (1618–48) stood out for their horrors, affecting both belligerents and non-combatants.

Writers of the following century thus saw it as enormous progress when states began to get a grip on taxation and became capable of paying their soldiers on a regular basis, thus forestalling the need for them to secure replacements for arrears in pay or to make up for redundancy after the end of a campaign by looting and worse. With some important exceptions, warfare in Europe became more 'regular', pitting regular armies against each other on battlefields, and leaving non-combatants relatively unscathed. Horrible exceptions included atrocities committed under Louis XIV against the inhabitants of the Palatinate, and in the Silesian and Seven Years Wars.[6] But on the whole, the world as Clausewitz had known it before the French Revolutionary Wars was relatively free of 'collateral damage', as we would say today. Before Clausewitz, General Count Guibert (1743–90) in France had remarked upon this;[7] as had Johann Friedrich von Decken, who opined that 'wars have become rarer and less destructive' thanks to the state monopolies of standing, regular armies.[8]

Even so, next to the 'regular' wars waged by these regular armies, another form of war existing side by side with it, or rather, on its periphery, and this was called, in the language of the martial culture that dominated Europe from the Grand Siècle until the mid-nineteenth

---

[5]Geoffrey Parker: *Europe in Crisis, 1598–1648* (London: Fontana 1979).

[6]For examples, see Martin Rink, *Vom Partheygänger zum Partisanen: die Konzeption des kleinen Krieges in Preussen, 1740–1813* (Frankfurt/Main: Peter Lang 1999), 127.

[7]Jacques-Antoine-Hippolyte de Guibert, 'Essai général de tactique' [1772], in Guibert, *Stratégiques*, Jean-Paul Charnay and Martine Burgos (eds.), (Paris: L'Herne 1977), 187f.

[8]Johann Friedrich von Decken, *Betrachtungen über das Verhältnis des Kriegsstandes zu dem Zwecke der Staaten* (Hanover: 1800, facsimile repr. Osnabrück: Biblio 1982), 134.

century, *la petite guerre*. It was not only practised much, but also much studied and analysed from the mid-eighteenth century, by soldier writers including Thomas-Antoine le Roy de Grandmaison, de Jeney, Count de la Roche, Roger Stevenson, Johann von Ewald, Andreas Emmerich, Georg Wilhelm Baron von Valentini, Gerhard von Scharnhorst, Friedrich Leopold Klipstein and Carl von Decker.[9] In Grandmaison we find the classical definition of the light forces (*parties* or *partis*) in the *petite guerre*:

> One can have several objectives in sending *parties* into combat. It may be to obtain news about the enemies, and to harass them whenever the occasion presents itself; it may be to seize a post or a detachment, to attack a convoy or stragglers, to seize baggage trains and foragers passing on horseback one after the other on their way to finding food; finally, to exact contributions [from the locals] or to carry out other missions far away.[10]

In the eighteenth century, Small Wars of the *petite guerre* type usually took the form of special operations carried out by special forces, namely small, light units or *détachements*, in German *Blänkerer* or *Plänkerer* (Scharnhorst), against the enemy's regular army or indeed

---

[9]Capt. Thomas-Antoine le Roy de Grandmaison, *La Petite Guerre, ou traité du service des troupes légères en campagne* (s.l., s.p. 1756); Capt. de Jeney, *Le partisan, ou l'art de faire la petite-guerre avec succes selon le génie de nos jours* (The Hague: Constapel 1759); Count de la Roche, *Essai sur la petite guerre; ou méthode de diriger les différentes opérations d'un corps de 2500 hommes de troupes légères* (Paris: Saillant & Nyon 1770); Roger Stevenson, *Military Instructions for Officers detached in the Field containing a scheme for forming a corps of partisans illustrated with plans of the manoeuvres necessary in carrying on the petite guerre* (London: John Millan, Edward & Charles Dilly 1770); Anon. (Johann Ewald), *Gedanken eines Hessischen Officiers über das, was man bey Führung eines Detaschements im Felde zu thun hat* (Cassel: Johann Jacob Cramer 1774) and idem, *Abhandlung von dem Dienst der leichten Truppe* (Flensburg/Schlwesig/Leipzig: 1790 and 1796); Andreas Emmerich, *The Partisan in War or the use of a corps of Light Troops to an Army* (London: H. Reynell for J. Debrett 1789); Gerhard von Scharnhorst, *Militärisches Taschenbuch zum Gebrauch im Felde* (Hanover: Helgsche Hofbuchhandlung 1793); Georg Wilhelm Baron von Valentini, *Abhandlung über den kleinen Krieg und über den Gebrauch der leichten Truppen mit Rücksicht auf den französischen Krieg* (1st ed. 1799, 2nd ed. Berlin: J.B. Boicke 1820); F.L. Klipstein, *Versuch einer Theorie des Dienstes der leichten Truppen, besonders in Bezug auf leichte Infanterie* (Darmstadt: Heyer 1799); Major Carl von Decker, *Der kleine Krieg im Geist der neueren Kriegsführung oder Abhandlung über die Verwendung und den Gebrauch aller drei Waffen im kleinen Kriege* (Berlin: Mittler 1822).

[10]Grandmaison, *La Petite Guerre*, 111.

his own light units.[11] Usage of the term 'Small War' itself goes back at least to the early seventeenth century, where it is found as in Spanish as *'guerrilla'* in 1611 in the dictionary of Sebastián de Covarrubias.[12] By the eighteenth century, the term is widely known, especially in French military literature, as *'petite guerre'*, or in English as 'Small War', and in German as *'der Kleine Krieg'*. This referred to small-scale military operations that were usually conducted by small 'detachments' of often irregular forces on the fringes of major operations by regular forces. The irregulars – or *'parties'* (French), *'Partheyen'* (German) were led by a 'partisan'. In the Universal Dictionary of Zedler of 1735, we find the following definition for *Parthey*:

> a military term, meaning a group of soldiers on horseback or on foot, sent forth by the general in order to harm the enemy through stratagems and speed, or to gather intelligence about him. .... The leader of such a party is called *Partheygänger* or partisan.[13]

'Partisan' thus originally referred only to the leader; gradually, it was applied also to the other members of the *'partie'/'Parthey'*, the *'Partheygänger'*. This is also how the terms were used in the earliest writings on the subject, by Grandmaison and Jeney in France, and by Andreas Emmerich, who fought for the Hanoverian ruler of Britain in the American War of Independence, leading light troops whom he referred to as 'partisans', and these were thus not hostile forces, but one's own special forces.[14] Their two purposes were thus to harass the enemy (*'Harcellierung'* in early modern German), or to gather intelligence. They were thus special forces, hired by a government or a general to work alongside regular forces. They were mercenaries, professional soldiers, often hailing from a tribe or ethnic group that specialised on this form of warfare, but they were not fighting for any ideology or even religious cause.[15]

They fought in small formations: estimates of what the best size should be differed. The Swiss Captain de Jeney thought a good size could be anything between 100 and 2,000 men, depending on

---

[11]Johannes Kunisch (ed.), *Gerhard von Scharnhorst: Private und dienstliche Schriften*, Vol. 1: *Kurhannover bis 1795* (Köln: Böhlau 2002), e.g. 161. The expression is used here also of the enemy's light troops.

[12]See Vittorio Scotti Douglas, 'Spagna 1808: la genesi della guerriglia moderna. 1. Guerra irregolare, 'petite guerre', 'guerrilla', *Spagna contemporanea* No. 18 (2000), 20.

[13]Zelder, *Universal-Lexikon* Vol. 26 (1740), Sp. 1049, quoted in Rink, *Vom Partheygänger*, 92.

[14]Emmerich, *Partisan War*, *passim*.

[15]Grandmaison, *La petite guerre*, 1–15.

circumstances, and for Jeney they should be a mix of light cavalry ('*dragons et hussards*') and infantry.[16] Johann Ewald, a Hessian who had fought in the Seven Years War and later on the British side in the American War of Independence, refers to these special forces as *Chasseurs* (German: *Jäger*, actually huntsmen, but here: riflemen); according to him, they were all-volunteer forces in the Seven Years War, configured into detachments of 500–600 men, while in America, they were configured into 'Light Corps' of around 1,000 men. Nevertheless, there might be missions where it could be useful to deploy only 30–40 infantrists.[17] For de la Roche, the total number of these light forces might be 2,500 on any one side.[18] Opinions also differed as to whether more mature men should be used in these parties, or boys between the ages of 16 and 18 in view of the hardships which older men would find it difficult to bear, as Ewald opined.[19]

They had to avoid pitched battles or other frontal encounters with regular forces, and their tactics differed significantly from those of regular forces. For example, instead of firing salvoes, they would employ *tirailleur* or sniper tactics, meaning that each would fire his weapon separately, aiming very carefully. They favoured terrain in which they could hide easily, in the marches, that is poorly accessible border areas, which were often mountainous or covered in forests, or swamps. Consequently, it was difficult to bring along heavy weapons, carriages or wagons of any sort, which meant that soldiers often had to dismount to fight, or that they could get around only on foot.

Clausewitz, in his notes on Small War, only wrote about light troops who were needed on one's own side, of *Hussars* and *Jäger*; in his script we also find the word *Tirailleur* but it is struck out.[20] By contrast, he referred to the adversary's forces as 'the enemy', and thus did *not* deal with ways of combating enemy irregular forces.[21] For Clausewitz, Small War was thus a special form of military operations, in which battalions of infantrists were deployed, sometimes also cavalry, in special 'detachments' in order to reconnoitre enemy positions. Clausewitz's notes thus focused especially on terrain and on precautions dictated by its nature, as in crossing ranges of hills or

---

[16]Jeney, *Le partisan*, 1f.

[17]Ewald, *Gedanken*, 74, and *Abhandlungen*, 18f.

[18]De la Roche, *Essai sur la petite guerre*.

[19]Ewald, *Abhandlungen*, 14.

[20]Clausewitz's lectures on Small War are printed in Werner Hahlweg (ed.), *Carl von Clausewitz: Schriften - Aufsätze - Studien - Briefe*, Vol. 1 (Göttingen: Vandenhoeck & Ruprecht 1966). See footnote on p.237.

[21]E.g. Ibid., 311.

mountains, rivers, swamps or moors, which simultaneously held dangers and offered opportunities.

Small War in Early Modern History, until Clausewitz's times, was thus really an auxiliary part of major war. The majority of authors, Clausewitz among them, assumed that Small War would be a side show to the main confrontations of regular forces. The encounters that would decide the war were expected to take place elsewhere, on open battlefields, between regular armies.

Clausewitz's little-known colleague at his General Military School, the later General Otto August Rühle von Lilienstern, a decade ahead of Clausewitz[22] recognised that even after Napoleon, war was not only major war, but that 'war in general is composed of both forms of war, or constituted jointly by both', and that 'the events and operations which take place in the course of a campaign belong, in part, to the domain of Small War, in part to the domain of Major War'.[23] It was for this reason, Rühle opined, that writers had such problems with the definition of Small War. On the whole, 'the domain of Small War was that of operations of smaller scale and employment of resources, while the deployment of the armed forces as a whole, and the decisive strikes, would be sought in the domain of Major War'. And yet he noted that Small Wars 'in certain circumstances could reach such a scope that several thousands [of combatants], indeed, the larger part of the existing and assembled forces, might have to be used for its purpose. Some authors', he continued, 'have seen Small War' as essentially only the 'operations of light forces, yet light forces are being used in our current style of warfare just as frequently in Major War ...'. And vice versa: in special circumstances, regular, heavily armed forces might be used in Small Wars. Rühle saw as a persistent difference that there was 'no or no significant' fighting in Small Wars. While 'the existence and the possibility of the success of the larger-scale operations' might depend directly on the success of the small-scale war-fighting accompanying it, the latter could only lead to any significant achievement, if its operations were closely coordinated with the large-scale operations, and worked in explicit support of the overarching endeavour.[24] Nevertheless, Rühle thought that his

---

[22]Clausewitz, 'Nachricht' vom 10. Juli 1827, in Werner Hahlweg (ed.), *Vom Kriege*, 19th ed. (Bonn: Dümmler 1991), 179–81.

[23][Otto August] R[ühle] v. L[ilienstern]: *Handbuch für den Officier zur Belehrung im Frieden und zum Gebrauch im Felde*, Vol. 2 (Berlin: Reimer 1818), 1; excerpts to be published in Beatrice Heuser (ed.), *The Strategy Makers: Thoughts on War and Society between Machiavelli and Clausewitz* (Westport, CT: Praeger/Greenwood forthcoming 2010).

[24]Rühle von Lilienstern, *Handbuch*, Vol. 2, p. 2f.

contemporaries did not pay enough attention to Small Wars defined in this way. This is true also for parts of Clausewitz's *On War*, which were extensively focused on Napoleon, whose 'entire desire had been focused on coercing his enemy, as quickly as possible, to give him the decisive main battle, ... for which reason neither [Napoleon] nor his enemies really had the time or the opportunity to deal much with Small War', as Rühle remarked.[25]

Having looked at definitions of Small War, let us cast a quick look at the real historical examples which Clausewitz would have known or read about in his studies of eighteenth century warfare. There were many examples here of light forces being used by armies for special tasks such as reconnaissance, but also for acts of sabotage, or small-scale harassment of enemy forces.[26] In Central and Eastern Europe it was common for such light forces to be recruited from special ethnic groups. The Habsburgs were spoilt for choice, as they had so many small ethnic groups specialising in horsemanship but also in certain techniques of warfare within their Empire. The most famous of these were the Habsburg Empire's Hussars, originally a formation of Hungarian cavalry dating back to 1688; in the following century, other states followed the Habsburgs' example of creating such special forces. Croats, Serbs, Bosniaks and Vlachs had similar skills and were also recruited for such tasks, in view of their special familiarity with mountain warfare. The Cossacks in Russia excelled in horsemanship, Tatars, Pandurs, Kalmyks and Turkic tribes likewise.

In addition, for the purpose of intelligence-gathering one needed individuals who knew many languages, and who knew their way around the areas where wars were being fought. As partisans had a vested interest in getting on well with the local populations, it was thus advantageous to recruit them from the conflict zone itself. Nevertheless, as Martin Rink has remarked in his trail-blazing study of the subject, for the partisans the fate of the local populations, often composed of many different ethnic groups living next to each other or in separate villages, was not the main concern; they were first and primarily mercenaries. There may have been some among them who occasionally imagined they fought for a cause, or who felt a particular loyalty to one prince rather than another, but ultimately, the majority among them did not fight for their own cause, nor even for their family or village or people, they fought, with loyalty to their leader, for money and for their band of soldiers.[27] In the wars of the *ancien régime*, they were most emphatically not freedom fighters or political agitators.

---

[25]Ibid., 3.

[26]Werner Hahlweg, *Guerilla: Krieg ohne Fronten* (Stuttgart: Kohlhammer 1968), 27f.

[27]Rink, *Vom Partheygänger*, 124.

It was this sort of use of force which was at the centre of Clausewitz's notes on Small War, written for a lecture series, held by him at the General War School in Berlin in 1810–11.[28] As a 30-year old lecturer with the rank of a major, Carl von Clausewitz built on the existing literature on the subject, illustrating them with examples from the recent past. Clausewitz explicitly referred to the lectures of his colleague Carl Ludwig Heinrich von Tiedemann (1777–1812), which, as the great Clausewitz-expert Werner Hahlweg has established, were circulated in manuscript form under the title 'Lectures about Tactics, edited by a general staff officer'.[29] Apart from this, Clausewitz referred directly to the works of Johann Ewald,[30] Andreas Emmerich[31] and Valentini,[32] as well as Archduke Charles,[33] repeatedly referring deferentially to the field manual written by his own teacher and mentor, the then Major General Gerhard von Scharnhorst.[34]

The absence of any reference to Grandmaison, de Jeney and La Roche is striking. While Clausewitz's hatred of the French language let alone all things French is well known, he could read French well and in other contexts drew on other French authors.[35] One would thus have assumed that these works were known to Clausewitz either directly or indirectly,

---

[28]Gottlieb Friedländer, *Die Königliche Allgemeine Kriegsschule und das höhere Militair- Bildungswesen 1765–1813* (Berlin: Mittler 1854).

[29]Hahlweg, *Carl von Clausewitz*, Vol. 1, footnote 2, 228.

[30]E.g. Hahlweg, *Carl von Clausewitz*, Vol. 1, 295, 351. These are Johannes Ewald, *Abhandlung über den kleinen Krieg* (Cassel: Johann Jacob Cramer 1785). Contains: 'Beytrag von dene drey vornehmsten Stücken was ein Officier von der leichten Reuterey im Feld zu tun hat'. Trans. into English: *Treatise on Partisan Warfare*, trans. Robert A Selig and David Curtis Skaggs (New York: Greenwood Press 1991); *Gedanken eines Hessischen Officiers über das, was man bey Führung eines Detaschements im Felde zu thun hat* (Cassel: Johann Jacob Cramer 1774); *Belehrungen über den Krieg, besonders über den kleinen Krieg, durch Beispiele großer Helden und kluger und tapferer Männer* (Schleswig: J.G. Röhss 1798); *Folge der Belehrungen über den Krieg, besonders über den kleinen Krieg, durch Beispiele grosser Helden und kluger und tapferer Männer* (Schleswig: J.G. Röhss 1800); *Abhandlung von dem Dienst der leichten Truppe* (Flensburg/ Schleswig/ Leipzig 1790 and 1796).

[31]E.g. Hahlweg, *Carl von Clausewitz*, Vol. 1, 355; Emmerich, *The Partisan*.

[32]E.g. Hahlweg, *Carl von Clausewitz*, Vol. 1, 304; Valentini, *Abhandlung über den kleinen Krieg*.

[33]E.g. Hahlweg, *Carl von Clausewitz*, Vol. 1, 304; Archduke Charles, *Grundsätze der höhern Kriegskunst für die Generäle der österreichischen Armee* (Vienna: Imperial & Royal Printing Office 1806).

[34]Hahlweg, *Clausewitz: Vom Kriege*, 335; Hahlweg, *Carl von Clausewitz*, Vol. 1, 349; Gerhard Scharnhorst, *Militärisches Taschenbuch zum Gebrauch im Felde* (1793, 2nd imp. 1793, 3rd imp. Hanover: Helwigsche Hofbuchhandlung 1794, 4th imp. 1815).

[35]On Clausewitz's views of the French, see his 'Bekenntnisdenkschrift', Hahlweg, *Carl von Clausewitz*, Vol. 1, 736.

not least as he himself had taken notes on Scharnhorst's lectures of 1804, who in turn was well aware of these books on Small War.[36]

What Clausewitz had to say on this subject in his lectures is neither original nor memorable in the light of subsequent developments and changes in tactics, technology, and the spirit of war. The only part perhaps worth quoting is Clausewitz's description of the quality which a partisan leader should have in special measure:

> Small warfare has the peculiar characteristic that it couples extreme boldness and courage with a much greater fear of danger than in major war. This characteristic is shared by the troops that conduct it. The individual Hussar or *Jäger* has an enterprising spirit, a degree of self-reliance and faith in his own luck which is almost unimaginable to somebody who has always served in line [among the regulars]. In the light of his experience and customs, he feels calm and unruffled while carrying out diverse and difficult missions which would make a [regular soldier] very anxious. By contrast, the Hussar and *Jäger* has a greater respect for the dangers of a regular battle than soldiers in regular formations. If it is not for dire necessity, he would not expose himself to it, but would retreat and seek cover as often as he can. A single cannon shot can keep troops at a great distance, if they do not carry cannon with them and [thus] feel disadvantaged. 100 infantrists could thus keep whole cavalry regiments at bay, if these saw the terrain of the infantrists as advantageous to these and perceived an asymmetry.

And he commented on the particular independence of irregular-forces:

> This is a necessary feature of the light troops. If this were not so, how would it be possible for them to carry on in full sight of the enemy, engaging in skirmishes almost every day without being wasted in one single campaign? One must not blame the light troops for these characteristics, by contrast, these are vital to them. Extreme boldness and wise prudence must take their turns and each individual must be equally good at both.[37]

Even today, this analysis fits small bands of fighters, whether they be the special forces that accompany the operations of regular armies behind the lines of the enemy, or whether they fight as what today we would call 'guerrillas' or 'partisans'.

---

[36]Johannes Kunisch (ed.), *Gerhard von Scharnhorst: Private und dienstliche Schriften*, Vol. 3: *Lehrer, Artillerist, Wegbereiter* (Köln: Böhlau 2005).
[37]Hahlweg, *Carl von Clausewitz*, Vol. 1, 237f.

Clausewitz did not see this sort of partisan warfare in an ideological context. He never focused on counter-insurgency. While he read Ewald's generalisations drawn from that officer's counter-insurgency experience in the American War of Independence, Clausewitz was not alerted to the factor of ideology which was creeping into this form of war. To Ewald by contrast it had been clear that counter-insurgence comprised the struggle for the hearts and minds of the population on whose territory this struggle was carried out. He emphasised the need to discipline one's soldiers lest they trouble the local populations; only if they were treated with great respect and forbearance could one hope to gain their respect and support in turn.[38]

The Second Manifestation of Small Wars: 'People's War' or Insurgency[39]

Both up to 1648 and again at the time of the French Revolutionary and Napoleonic Wars, conflicts had been common in which, on one side or the other, the civilian population had been greatly affected, either as victims of repressions or forced requisitioning by occupying powers, or, often in reaction to such treatment, as insurgents against occupying forces. The intervening years had seen comparatively little of the sort. Then the American War of Independence of 1775–83 was the first in a new series of such insurgencies, even though in part it had the character of classic regular war with pitched battles between regular forces. On the side of the rebels these were complemented, however, by irregular forces, such as the ones Ewald and Emmerich had described in their books. On the American side, both the regular and the irregular forces were made up mainly of 'militias', namely citizen-soldiers defending their own land. In the seventeenth and eighteenth centuries, this term was used synonymously with 'military' or any members of the armed forces, or else in the sense of locally recruited soldiers as opposed to professional forces or mercenaries.[40] The American militias generally had difficulties defeating the British redcoats in conventional battles, but they had all advantages on their side when harassing the redcoats on the move, especially when they had to cross forests or swamps. General George Washington's regular forces were supported

---

[38]Ewald, *Abhandlung*, 16.
[39]The contemporary terminology does not distinguish in as clear-cut a way as I am trying to. Le Mière de Corvey called the Spanish insurgents *'parties de guerrilla'*; *Partheygängerkrieg* is used by Chrzanowski. This is why Clausewitz's differentiation, referring to irregulars fighting alongside regular armies as 'partisans' *(Partheygänger)* and the insurgency against occupation forces 'people's war' is particularly helpful.
[40]Rink, *Vom Partheygänger*.

by irregulars who made considerable contributions to the defeat of the British.

Clausewitz was familiar with the American War of Independence, and explicitly referred to Washington's larger battles.[41] As Clausewitz referred to books on this subject in his lectures, like Emmerich's (see above), he must have been aware of it, and yet he made no explicit reference to the unconventional or indeed ideological element of this war.

The phenomenon of the armed insurgency probably goes back as far in history as any form of war, probably to times before any formal distinction was made between combatants and non-combatants. The French Revolution drew on certain characteristics of the armed mass-insurgency of untrained people with the *levée en masse* of 1793, with the famous and often quoted passage of total mobilisation, assigning roles to all sectors of the population, from the very young to the very old.[42]

While this expressed the political ideal of the active support of the Revolution by the entire population, what happened in fighting is that the undisciplined and untrained soldiers of the *levée en masse* applied Small War tactics to large-scale encounters. Instead of shooting volleys and marching forward row by row, each picked his enemy to target, and apart from a number of large-scale battles, the French harassed enemy forces wherever they could find them. Thus the Austrian Archduke Charles, who got a taste of the new French method, described the French 'people's war' as a form of war where the point was not 'to counter enemy operations frontally, but to trouble and affect them through diversions'.[43] Another eyewitness, Julius von Voss, wrote in 1809 about French tactics that 'now all is Small War on a large scale'.[44]

As a counter-revolutionary and a monarchist, Clausewitz sympathised with the insurgency of the population of the Vendée in their reactionary resistance to the French Revolution, 1793–96, as it was a monarchist rebellion against the new regime and the new ideology. This

---

[41]Hahlweg, *Carl von Clausewitz*, Vol. 1, 321 'General Wassington' [*sic*]. With regard to Washington's strategy of attrition, see Russell F. Weigley, *The American Way of War* (New York: Macmillan 1973), 3–17.

[42]Quoted in William Serman and Jean-Paul Bertraud, *Nouvelle Historie Militaire de la France* (Paris: Fayard 1998), 75f.; see also Jean Delmas (ed.), *Histoire Militaire de la France, Vol. 2: 1715–1871* (Paris: Presses Universitaires de la France 1992), 241–5.

[43]Archduke Charles, 'Das Kriegswesen in Folge der französischen Revolutionskriege' (1838), in Baron von Waldtstätten (ed.), *Erzherzog Karl - Ausgewählte militärische Schriften* (Berlin: Richard Wilhelmi 1882), 209.

[44]Julius von Voss, *Der Kleine Krieg* (1809), preface, p. iii., quoted in Rink, *Vom Partheygänger*, 197.

insurgency became one of his two great models for Prussian resistance against Napoleon. It seems that Clausewitz had read Alphonse de Beauchamp's *Histoire de la guerre de la Vendée et des Chouans, depuis son origine jusqu'à la pacification de 1800,* which was first published in three volumes in 1806 and reprinted twice in the 1807 and 1809.[45] From this he concluded that much could be done by even minimally armed peasants in a popular uprising. Against the argument that the Vendéeans had special geographic advantages on their side in Poitou and Anjou, and that only this had enabled them to hold out so long, he replied that 'the swampy forests of Pomerania, East Prussia, West Prussia and Silesia' would be a perfect match for these, and lent themselves beautifully to an insurgency.[46]

Another example Clausewitz cited was that of the popular insurgency in the Tyrol in 1809, again an event that occurred near the time of his writing.[47] After the Austrian Army had been beaten at the Battle of Wagram on 5–6 July 1809 and the Habsburg Archduke Charles concluded an armistice with the French at Znaym on 12 July 1809, the Tyroleans refused to lay down their arms and to accept the rule of the Bavarians, Napoleon's allies. They, too, staged a popular uprising. French troops under Marshal Lefebvre tried to quash it, but without success, as typically for this type of warfare, the Tyroleans retreated into the mountains, eluding the Bavarian and French forces who set out to catch the rebels. Originally, Austrian chief minister Klemens Metternich had tried to support the rebels, but ceased to do so when he realised that they were escaping his control. Archduke Charles saw this uprising 'as the *last* desperate resort, but *only* to the point as one has the hope of bringing about better circumstances or to continue with it until these materialise'. For in general, he feared the 'disadvantage' which this kind of war inevitably entailed, namely 'that it undermines the prosperity and the virtues of the people for a long time'.[48] Like the Prussian government, Archduke Charles was extremely worried that a people's war might lead to demands for social and political emancipation.

The most famous popular insurrection of this type of Small War, the *Guerrilla* in Spain 1808–14, would in the minds of later generations permanently change the connotation of the Spanish term. In 1808 Spaniards rose up spontaneously against Napoleon and his armies,

---

[45]Carl von Clausewitz, 'Bekenntnisdenkschrift von 1812', in Hahlweg, *Carl von Clausewitz*, Vol. 1, 721 and footnotes 11 and 134.

[46]Ibid., 721, 732, 734.

[47]E.g. Hahlweg, *Carl von Clausewitz*, Vol. 1, 311, 721.

[48]Archduke Charles, 'Geist des Kriegswesens überhaupt', c.1823–26, in Waldstätten, *Erzherzog Karl*, 110f.

when Napoleon deposed the Spanish King Charles IV and his son Ferdinand VII, replacing them with his own brother, Joseph Bonaparte. The self-appointed Spanish government, the *Junta*, which claimed to be acting for the deposed Ferdinand VII, legitimated the insurgency, but the insurgents in large part felt they were fighting for their own cause, for which the king was an important symbol, but merely a symbol.

Contrary to long nurtured myths about the Spanish *Guerrilla*, it was only in part fought by peasants, monks and women brandishing pitchforks. Instead, physical opposition to the French occupation came from three organised indigenous groups, the demarcations between whom were fluid, plus the British and Portuguese forces engaged in the Peninsular War. There were indeed local insurgents, engaging sporadically in more or less spontaneous violence, and they could be described as the first group.[49] But more of the fighting was carried out by a second category of Spaniards, irregular troops, recruited mainly from the local population, but soon subjected to military drill, fighting under experienced military leaders who in turn were under the strict control of the Junta. These volunteers (*volunarios*), Small War troops (*la tropa de Guerrilla*),[50] *cuadrillas, quadrillas,* Free Corps (*cuerpos francos*) or parties (*partidas*)[51] soon sported uniforms, and behaved very much like the partisans or light special forces of the *ancien régime*. They differed from these earlier partisans, however, in that the Spaniards were defending their own property, their own families and village communities. Most of them fought in defence of their own homeland against the foreign occupants, which introduced the crucial new dimension of political cause and ideology.[52] Like the partisans of old, however, the Spanish *Partidas* also had economic incentives to fight against the French: they either received regular pay (more than regular soldiers!), or else they ensured that they had revenues through punishment actions against real or supposed collaborators with the

---

[49]Charles Esdaile, *Fighting Napoleon: Guerrillas, Bandits and Adventurers in Spain, 1808–1814* (New Haven, CT: Yale UP 2004), 43 f.

[50]Don Gonzales Ofarrill, *Instruccion que deben seguir los oficiales y tropas del 1er batallon de voluntarios de Cataluña quando se empleen en guerrilla, o como tiradores* (Liorna: Imprenta de Antonio Vignozzi 1806).

[51]José Maria de Carvajal, *Reglamento para las Partidas de Guerrilla* (Cádiz Don Nicolas Gomez de Requena 1812).

[52]It does not help our purpose of seeking terminological clarity that later in the nineteenth century, the Spaniards took to employing the term *guerrilleros* as a synonym of *tiradores*, in the eighteenth century sense of 'partisan', i.e. special forces or detachments, particularly employed as snipers, it seems, employed like regular soldiers directly by the government, and subject to considerable drill – see Anon., *Instruccion de guerrilla* (Spoleto: Tipografia Bossi & Bassoni 1849) and Gen. Don Fernando Fernández de Córdova, *Tactica de guerrilla* (Madrid: s.p. 1870).

French.[53] In total, there may at the height of the *Guerrilla* have been between 40,000 and 50,000 such irregular fighters.[54] According to Le Mière de Corvey, these irregulars fought in units of about 120–180 men,[55] but there were occasions when they grouped into much larger formations, partly alongside regular forces, and actually delivered real battles (albeit usually with little success).[56]

The third group of fighting Spaniards was made up of the regular Royal Army. The division between these (otherwise not very successful) regular forces and the partisans was fluid.[57] The operations of the regular armies and the *partidas* were often indistinguishable. Some partisan leaders or *cabecillas* like Sánchez were or had been officers in the regular armies; others had been anything from landowners, innkeepers or professionals down to bandits or the odd monk.[58]

The *Guerrilleros* like the soldiers of the French Revolutionary armies in large part applied *ancien régime* Small War tactics, but the extent of the uprising, the involvement of untrained soldiers and their own cause that they were fighting for gave the whole enterprise a different, larger dimension. As a Prussian participant in these wars, Heinrich von Brandt, conceded, even the *Guerrilla* was ultimately successful only because Napoleon was decisively defeated in conventional battles elsewhere, and that it continued to be, as in the eighteenth century, part of a larger effort.[59] Nevertheless, the *Guerrilla* had the dimension of a popular uprising or insurgency, with a political cause, which the Small Wars of the mercenary partisans of the *ancien régime* had lacked.

Both the Russians and the Prussians purported to be following the Spanish example in their resistance to the *Grande Armée* in 1812 and 1813 respectively.[60] If we follow Charles Esdaile's study of the socio-economic underpinning of the *Guerrilla*, Clausewitz was

---

[53]Esdaile, *Fighting Napoleon*, 38f; Vittorio Scott-Douglas, 'Regulating the Irregulars: Spanish Legislation on *la guerrilla* during the Peninsular War', in Charles J. Esdaile (ed.), *Popular Resistance in the French Wars: Patriots, Partisans and Land Pirates* (Basingstoke, UK: Palgrave Macmillan 2005).

[54]Charles Esdaile and Leonor Hernández Enviz, 'The Anatomy of a Research Project: The Sociology of the Guerrilla War in Spain 1808–14', in Esdaile, *Popular Resistance in the French Wars*, 120.

[55]J.F.A. Le Mière de Corvey: *Des Partisans et des corps irréguliers* (Paris: Anselin & Pochard 1823), 100.

[56]Esdaile, *Fighting Napoleon*, 45.

[57]Ibid., 44–50, 53–7.

[58]Ibid., 93.

[59]Heinrich von Brandt, *Ueber Spanien mit besonderer Hinsicht auf einen etwaigen Krieg* (Berlin: Schüppel'sche Buchhandlung 1823), 57ff.; published in English as *The Two Minas and the Spanish Guerrillas* (London 1825).

[60]John Ellis, *A Short History of Guerrilla Warfare* (London: Ian Allan 1975), 66–9.

wrong, at least where the Spanish Guerrilla was concerned, when assuming that the insurgents were cheaper than mercenaries.[61] For Prussian purposes, Clausewitz counted on the idealism of his putative insurgents:

> An army that fights on its own soil and ground for its highest interests cannot and must not be regarded as a bunch of mercenaries who carry their own skin to market in return for cash; instead, extreme abstinence from all luxury and plenty must be the spirit of such an army. In that case, we do not need the high salaries which are currently thought necessary to reward long service or retain talent. The fight for the Fatherland is the most beautiful reward for merit, the greatest attraction for talent.[62]

Clausewitz commented on the still ongoing Spanish *Guerrilla* in a treatise that he wrote in February 1812, known as his *Bekenntnis-denkschrift* or memorandum in which he professed his – at the time politically unacceptable – views. This he sent to his mentor, the then Major General August Neidhardt von Gneisenau, unofficial leader of the anti-French party in the Prussian officer corps and severe critic of the Prussian king's compliance with Napoleon's wishes in the treaty of 24 February 1812.[63] With his memorandum, Clausewitz threw in his lot with the reformers, and then himself quit Prussian military service to join the Russians who refused to give in to France. In the treatise, Clausewitz explicitly described the *Guerrilla* as a model for a Prussian resistance against the French, as we shall see presently.[64] Clausewitz noted that the French had:

> half of their entire manpower in Spain, namely 300,000 men ... while they fought their main battles against Wellington with armies of [only] 40,000 and 50,000 men; all the others were deployed to keep the insurgent troops apart and to prevent the uniting of all insurgent forces.[65]

Word had spread that the *Guerrilla* went along with unspeakable atrocities committed by both sides which Clausewitz did not wish upon

---

[61]Esdaile, *Fighting Napoleon*, 94–101
[62]Hahlweg, *Carl von Clausewitz*, Vol. 1, 711.
[63]Ibid., 732, 734.
[64]Ibid., 729.
[65]Ibid., 732.

his country; he gave some consideration to the question how such developments could be avoided. He thought that:

> these extremes, about which one is hearing in Spain, do not necessarily have to occur everywhere, and could perhaps be avoided by the measure alone that the government takes each armed man under its authority, and threatens reprisals against prisoners for each atrocity which is carried out [by the enemy] against the laws and customs of war against these real defenders of the Fatherland. How many executions [of his own men] will the enemy tolerate? And what are a few dozen people who are ready to die in this way for their Fatherland, against the mass of victims, which war claims on a daily basis?

> Indeed, the images one has of this danger – not more of a danger than any other in times of war – are greatly exaggerated. Even in Spain things are not as bad as we hear, and the enemy would be persuaded, after a few shots across the bows by his military police, to treat the insurgent troops like any others.[66]

The Prussian officer thus continued to advocate a popular insurgency. There were people, he wrote:

> who tremble at the thought of a people's war, because it is bloodier than others, is rarely free of horrible scenes, and all misery and destruction is multiplied in it. But whose fault is this? The people's war exists, you curse its pernicious effects, so curse those who have forced it upon us. If you make yourselves the judges of human actions, do not condemn the oppressed because he is weak. But be just, cast your curses against him who has made this evil necessary![67]

Gneisenau, Clausewitz and others had clearly been mulling over all the options, and Clausewitz was familiar with all the counter-arguments. For the politically conservative and undemocratic Prussian military leadership, the thought of politicised masses as the French Revolution had produced, or indeed the counter-revolutions of the Vendée and Spain, was a nightmare. A people's war, or '*Landsturm*' (another term used by Clausewitz for this phenomenon, meaning home guard or defensive militia) would pose a 'threat' to the 'existing order of things',

---

[66]Hahlweg, *Carl von Clausewitz*, Vol. 1, 734.
[67]Ibid., 740f.

which could indeed 'arise from this to the very governments, which try to apply this measure'. To this argument Clausewitz replied:

> This is precisely where our situation has its advantages: the government which provokes this storm remains its master. It is able to give it a general direction, and to drive it towards one goal. Even the divergence in opinion and action which in Spain is visibly destroying a large part of the effects, and which before divided the forces of the Vendée, can and will be prevented by a government which behaves [as well] towards its people as does that of Prussia.[68]

What, then, did Clausewitz have in mind exactly for the people's war he wanted Prussia to fight against France? The Prussian alliance of 1812 with France in his view had imposed upon Prussia conditions no better than those an occupied country had to suffer. For him, the alternative was:

> the organisation of a military state which would be horrible for the French Emperor; the introduction of conscription for all..., the reorganisation of the army in a new spirit; the removal of the old weak and bad generals and staff officers; the acquisition of plenty of arms and munitions; the creation of safe depots in defended camps; the creation of a general militia; the punishment of mistakes that have been made and of unpatriotic attitudes among military and civilians; the rolling renewal of the army after 2–3 years of service, which by now after four years would have given us at least 150,000 trained men, and which would have made military service less burdensome to the subject.[69]

Clausewitz calculated that with Prussia, a population of 4.6 million souls, should have 450,000 men between the ages of 18 and 30, plus another 300,000 aged 30–40, making 750,000 in all. If only one in seven were recruited to a regular army, Prussia could muster 100,000 regular soldiers.[70] The other able-bodied men could complement the regular army by forming a 'Landsturm':

> Landsturm is what we might call a general arming of an entire people for the direct defence of a country. ... These are its

---

[68]Ibid.
[69]'Bekenntnisdenkschrift', Hahlweg, Carl von Clausewitz, Vol. 1, 700.
[70]Ibid., 708.

particulars: all able-bodied men between 17 and 59 who do not serve as regular soldiers are armed and belong to the *Landsturm*. The arms and equipment are nothing more than a rifle, or if that is not available, a pike or sickle, a backpack to carry a few days' provisions and some munitions, and a cap cushioned with straw to protect against strikes, and which at the same time carries a badge of the province and the local community.[71]

The Prussian officer explained his design, giving a concrete example of the organisation of a *Landsturm* for the areas to the west of the Prussian capital, Berlin (which included his native town Burg near Magdeburg), to create impediments for an enemy advancing from West to East:

As soon as the enemy advances, the alarm bells are rung, and the *Landsturm* of all parishes gathers: wherever it encounters enemy columns, it retreats laterally into the forests by several marches, avoiding battle.

As soon as the main body of the enemy columns has marched past, the *Landsturm* attacks the detachments and rearguard of wagons which follow, disperses them as much as possible, and retreats again, as soon as the enemy turns about [to counter-attack].

In this way, the march of the [enemy] columns on their way to meet up in Berlin is made more difficult, and the enemy soldiers get a foretaste, an inkling of what awaits him, namely a Spanish Civil War in Germany.[72]

The enemy, for Clausewitz, was the French, and it did not matter to him whether Prussia was dealing with the French armies of the *levée en masse* of 1793 (themselves the product of a democratic-republican inspired people's uprising), or the mass armies of Napoleon (who were inspired by French-nationalist imperialism and the cult of Napoleon, yet integrating soldiers of many ethnic origins). For Clausewitz, the only difference between the Revolutionary Wars and the Napoleonic Wars was the degree of perfection which the latter had reached. But Clausewitz recognised that both were at the top end of popular mobilisation – he understood that the French Revolution, as well as Napoleon, as well as the Spanish and Tyrolean reactions were wars that had become 'the cause of the people', and that were therefore fought all

---

[71]Ibid., 721.
[72]'Bekenntnisdenkschrift', Hahlweg, *Carl von Clausewitz*, Vol. 1, 729.

the more vehemently.[73] The enthusiastic mass armies were thus only the extreme expression of the fanatical terrorist or the small band of *guerrilleros* writ large. While he did not fully spell this out in *On War*, this becomes apparent in his confession-memorandum.

As the latter, Clausewitz's main piece of work on a popular insurgency, was written before he joined the Tsar's Imperial Russian Army, and before Napoleon made his ill-fated campaign in Russia, we naturally find no reference to the extensive harassment of the *Grande Armée* by Russian irregulars which so greatly contributed to its spectacular losses. But this was subsequently witnessed by Clausewitz at first hand, and analysed in his detailed study of the campaign.[74] This again, while not referred to explicitly in *On War*, forms part of the backdrop of Book VI on 'People's War', which concludes our survey of historical examples that Clausewitz knew of or witnessed indirectly in his own times.

Conservative Progressives and Conservative Reactionaries

The Prussian government was reluctant to embrace, let alone apply, the ideas and plans of Clausewitz and Gneisenau in early 1812. They did not share Clausewitz's conviction that the Prussian people could be controlled by their own loyalty to their benevolent rulers. Instead, they feared the dynamics if the people's uprising would fully unfold. They much preferred to retain as much control and impose as much discipline as possible to keep the initiatives of spontaneously formed bands of freedom fighters in check. Thus the small bands of partisans were dissolved again as quickly as possible, and mobilised male civilians were subordinated to the local government servants.[75]

Clausewitz's perspective was that of the would-be insurgent. In writing about popular uprisings against governments or occupying forces, Clausewitz used the term 'the insurgents' (*Insurgenten*), but how to counter them was not a subject to which he gave much attention, notwithstanding the fact that in 1831, he died of cholera during a campaign to quash a Polish insurgency against Russian and Prussian occupation.[76]

Like Clausewitz, and at much the same time as him, some other East European strategists took a great interest in the experience of the

---

[73]Clausewitz, *Vom Kriege*, Book VIII, Ch. 3B, 970f.

[74]Carl von Clausewitz, ' Feldzug von 1812 in Russland', in Werner Hahlweg (ed.), *Carl von Clausewitz: Schriften, Aufsätze, Studien, Briefe*, Vol. 2, Pt. 2 (Göttingen: Vandenhoeck & Ruprecht 1990), 717–924.

[75]Rink, *Vom Partheygänger*, 328, 371.

[76]E.g. Hahlweg, *Carl von Clausewitz*, Vol. 1, 311.

Spanish *Guerrilla*. Even though Carl von Decker's book on Small Wars, published in 1822, still stands in the tradition of Grandmaison and Jeney, his putative partisans were fighting an ideological war. Von Decker may have taught at the same place as Clausewitz, so it is highly likely that Clausewitz had read his book.[77] With his book published in 1823, Heinrich von Brandt, who had fought alongside the insurgents in Spain, also took an interest in how to apply the lessons of the *Guerrilla* to a defensive insurgency strategy for his own country.[78]

In his publication of the same year, the Frenchman Lieutenant Colonel Le Mière de Corvey drew on the Vendée and on his own experience in Spain to write a handbook on how to conduct an insurgency and how to deny one's land to an invading army.[79] The Polish General Wojciech Chrzanowski wrote a manual in 1839, very much based on the same ideas as those of Le Mière de Corvey and Clausewitz, in which he described how to deny the conquest of one's land to invading forces, by staging a people's war.[80]

But the other side also drew lessons from the *Guerilla*. Showing more concern for the interests of the occupation forces, Marshal Thomas Robert Bugeaud de la Piconnerie, Duc d'Isly (1784–1849), drew on the experiences of the Napoleonic forces in their (ultimately unsuccessful) counter-insurgency efforts in Spain in his own campaign in the French colonial possessions in North Africa.[81] At the time, whether somebody was inclined to see this conflict from the point of view of the insurgents or of the counter-insurgency forces was not determined by left-wing (republican, democratic) or right-wing leanings. It was determined instead by the fact or likelihood of one's own country being subject to occupation, and to patriotic or nationalist and perhaps conservative-reactionary concerns to prepare in advance a resistance to this. It is thus ironic that Chrzanowski's manual, which clearly had resistance against the Prussian and Russian occupation of Poland in mind, shows so many similarities to Clausewitz's People's War (with French occupation in mind) and Le Mière de Corvey's writing. Insurgency and counter-insurgency, at the time, could not be fitted into a political spectrum in the way in which it was in the Cold War.

---

[77]Decker, *Der Kleine Krieg*; Peter Paret, *Clausewitz and the State* (Princeton UP 1979), 316f.

[78]Brandt, *Ueber Spanien*.

[79]Le Mière de Corvey, *Des Partisans*.

[80]Gen. Wojciech Chrzanowski, *Uber den Parteigaenger-Krieg: Eine Skizze*, trans. from Polish (Berlin: Stuhrsche Buchhandlung 1846).

[81]Thomas Robert Bugeaud de la Piconnerie, Duc d'Isly, *De la stratégie, de la tactique, des retraites et du passage des défilés dans les montagnes des Kabyles* (Algiers: Imprimerie du Gouvernement 1850).

Unless one takes 'nationalism' to be the force of progress in Clausewitz's age (which in the Spanish case it clearly was not), the political spectrum was such that 'People's War' could be a phenomenon of the Left and the Right. No wonder it took so long for ideology to be fully recognised by contemporaries as the factor that had perhaps most profoundly transformed warfare since 1775 and 1789. Clausewitz's unconscious and yet so insightful and helpful use of two different terms – partisans' Small Wars versus People's War – was not recognised by him or his contemporaries as representing a watershed in the history of Small Wars. We should be careful thus to distinguish between retrospective analysis and the way Clausewitz and his contemporaries actually perceived the world.

Bibliography

Anon. (Johann Ewald), *Gedanken eines Hessischen Offcviers über das, was men bey Führung eines Detaschements im Felde zu thun hat* (Cassel: Johann Jacob Cramer 1774).

Anon., *Instruccion de guerrilla* (Spoleto: Tipografia Bossi & Bassoni 1849).

Bassford, Christopher, 'In defence of Clausewitz', *Times Literary Supplement*, 15 Jan. 1993.

Beckmann, Erasmus: *Clausewitz, Terrorismus und die NATO-Antiterrorstrategie*, AIPA 3/2008 *Arbeitspapiere zur Internationalen Politik und Aussenpolitik* (Cologne Univ. 2008).

Brandt, Heinrich von, *Ueber Spanien mit besonderer Hinsicht auf einen etwaigen Krieg* (Berlin: Schüppel'sche Buchhandlung 1823), published in English as *The Two Minas and the Spanish Guerrillas* (London: 1825).

Bugeaud de la Piconnerie, Thomas Robert, Duc d'Isly, *De la stratégie, de la tactique, des retraites et du passage des défilés dans les montagnes des Kabyles* (Algiers: Imprimerie du Gouvernement 1850).

Carvajal, José Maria de, *Reglamento para las Partidas de Guerrilla* (Cádiz: Don Nicolas Gomez de Requena 1812).

Charles, Archduke of Austria, *Grundsätze der höhern Kriegskunst für die Generäle der österreichischen Armee* (Vienna: Imperial & Royal Printing Office 1806).

Chrzanowski, Gen. Wojciech: *Uber den Partheigaenger-Krieg: Eine Skizze*, trans. from Polish (Berlin: Stuhrsche Buchhandlung 1846).

Córdova, Gen. Don Fernando Fernández de, *Tactica de guerrilla* (Madrid: s.p. 1870).

Daase, Christopher, 'Clausewitz and Small Wars', in Hew Strachan and Andreas Herberg-Rothe (eds.), *Clausewitz in the Twenty-First Century* (Oxford: OUP 2007), 182–195.

Decken, Johann Friedrich von, *Betrachtungen über das Verhältnis des Kriegsstandes zu dem Zwecke der Staaten* (Hannover: 1800, facsimile print Osnabrück: Biblio 1982).

Decker, Major Carl von, *Der kleine Krieg im Geist der neueren Kriegsführung oder Abhandlung über die Verwendung und den Gebrauch aller drei Waffen im kleinen Kriege* (Berlin: Mittler 1822).

Delmas, Jean (ed.), *Histoire Militaire de la France*, Vol. 2 *1715–1871* (Paris: Presses Universitaires de la France 1992).

Douglas, Vittorio Scotti, 'Spagna 1808: la genesi della guerriglia moderna. 1. Guerra irregolare, "petite guerre"', "guerrilla"', *Spagna contemporanea* No. 18 (2000), 9–32.

Echevarria III, Antulio, *Clausewitz and Contemporary War* (Oxford: OUP 2007).

Ellis, John, *A Short History of Guerrilla Warfare* (London: Ian Allan 1975).

Emmerich, Andreas, *The Partisan in War or the use of a corps of Light Troops to an Army* (London: H. Reynell for J. Debrett 1789).

Esdaile, Charles J., *Fighting Napoleon: Guerrillas, Bandits and Adventurers in Spain, 1808–1814* (New Haven, CT: Yale UP 2004).

Esdaile, Charles J. (ed.), *Popular Resistance in the French Wars: Patriots, Partisans and Land Pirates* (Basingstoke, UK: Palgrave Macmillan 2005).

Ewald, Johann, *Gedanken eines Hessischen Officiers über das, was man bey Führung eines Detaschements im Felde zu thun hat* (Cassel: Johann Jacob Cramer 1774).

Ewald, Johann, *Abhandlung über den kleinen Krieg* (Cassel: Johann Jacob Cramer 1785).Contains: 'Beytrag von dene drey vornehmsten Stücken was ein Officier von der leichten Reuterey im Feld zu tun hat'. Trans. into English: *Treatise on Partisan Wafare*, trans. Robert A Selig and David Curtis Skaggs (New York: Greenwood Press 1991).

Ewald, Johann, *Abhandlung von dem Dienst der leichten Truppe* (Flensburg/Schleswig/Leipzig: 1790 and 1796).

Ewald, Johann, *Belehrungen über den Krieg, besonders über den kleinen Krieg, durch Beispiele grosser Helden und kluger und tapferer Männer* (Schleswig: J.G. Röhss 1798).

Ewald, Johann, *Folge der Belehrungen über den Krieg, besonders über den kleinen Krieg, durch Beispiele grosser Helden und kluger und tapferer Männer* (Schleswig: J.G. Röhss 1800).

Friedländer, Gottlieb, *Die Königliche Allgemeine Kriegsschule und das höhere Militair-Bildungswesen 1765–1813* (Berlin: Mittler 1854).

Grandmaison, Capt. Thomas-Antoine le Roy de, *La Petite Guerre, ou traité du service des troupes légères en campagne* (s.l., s.p. 1756).

Gray, Colin S., 'Clausewitz Rules, OK? The Future is the Past – with GPS', *Review of International Studies* 25 (1999), 161–82.

Guibert, Jacques-Antoine-Hippolyte de, 'Essai général de tactique' [1772], in Jean-Paul Charnay and Martine Burgos (eds), *Guibert Stratégiques* (Paris: L'Herne 1977).

Hahlweg, Werner (ed.), *Carl von Clausewitz: Schriften - Aufsätze - Studien - Briefe*, Vol. 1 (Göttingen: Vandenhoeck & Ruprecht 1966).

Hahlweg, Werner, *Guerilla: Krieg ohne Fronten* (Stuttgart: Kohlhammer 1968).

Hahlweg, Werner, 'Clausewitz and Guerrilla Warfare', in Michael I. Handel (ed.), *Clausewitz and Modern Strategy* (London: Frank Cass 1986), 127–33.

Hahlweg, Werner (ed.), *Carl von Clausewitz: Schriften, Aufsätze, Studien, Briefe*, Vol. 2, Pt 2 (Göttingen: Vandenhoeck & Ruprecht 1990).

Hahlweg, Werner (ed.), *Clausewitz: Vom Kriege*, 19th ed. (Bonn: Dümmler 1991).

Heuser, Beatrice, *Reading Clausewitz* (London: Pimlico 2001).

Heuser, Beatrice (ed.), *The Strategy Makers: Thoughts on War and Society between Machiavelli and Clausewitz* (Westport, CT: Praeger/Greenwood forthcoming 2010).

Heuser, Beatrice, *Thinking War: The Evolution of Strategy from Antiquity to the Present* (Cambridge: CUP forthcoming 2010).

Hughes, Paul D., *Small Wars, or Peace Enforcement According to Clausewitz* (Carlisle Barracks, PA: US Army War College Strategic Studies Institute 1996).

Jeney, Capt. de, *Le partisan, ou l'art de faire la petite-guerre avec succes selon le génie de nos jours* (The Hague: Constapel 1759).

Kinross, Stuart, 'Clausewitz and Low-Intensity Conflict', *Journal of Strategic Studies* 27/1 (March 2004), 35–58.

Klipstein, F.L., *Versuch einer Theorie des Dienstes der leichten Truppen, besonders in Bezug auf leichte Infanterie* (Darmstadt: Heyer 1799).

Kunisch, Johannes (ed.), *Gerhard von Scharnhorst: Private und dienstliche Schriften*, Vol. 1: *Kurhannover bis 1795* (Köln: Böhlau 2002).

Kunisch, Johannes (ed.), *Gerhard von Scharnhorst: Private und dienstliche Schriften*, Vol. 3: *Lehrer, Artillerist, Wegbereiter* (Köln: Böhlau 2005).

Mière de Corvey, J.F.A. Le, *Des Partisans et des corps irréguliers* (Paris: Anselin & Pochard 1823).

Münkler, Herfried, *Über den Krieg: Stationen der Kriegsgeschichte im Spiegel ihrer theoretischen Reflexion* (Weilerswist: Velbrück Wissenschaft 2008).

Ofarrill, Don Gonzales, *Instruccion que deben seguir los oficiales y tropas del 1er batallon de voluntarios de Cataluña quando se empleen en guerrilla, o como tiradores* (Liorna: Imprenta de Antonio Vignozzi 1806).

Paret, Peter, *Clausewitz and the State* (Princeton UP 1979).

Parker, Geoffrey, *Europe in Crisis, 1598–1648* (London: Fontana 1979).

R[ühle] v. L[ilienstern], [Otto August], *Handbuch für den Officier zur Belehrung im Frieden und zum Gebrauch im Felde*. Vol. 2 (Berlin: Reimer 1818).

Rink, Martin, *Vom Partheygänger zum Partisanen: die Konzeption des kleinen Krieges in Preussen, 1740–1813* (Frankfurt/Main: Peter Lang 1999).

Roche, Count de la, *Essai sur la petite guerre; ou méthode de diriger les différentes opérations d'un corps de 2500 hommes de troupes légères* (Paris: Saillant & Nyon 1770).

Scharnhorst, Gerhard von, *Militärisches Taschenbuch zum Gebrauch im Felde* (Hanover: Helgsche Hofbuchhandlung 1793).

Scharnhorst, Gerhard von, *Militärisches Taschenbuch zum Gebrauch im Felde* (1793, 2nd imp.1793, 3rd imp. Hanover: Helwigsche Hofbuchhandlung 1794, 4th imp. 1815).

Serman, William and Bertraud, Jean-Paul, *Nouvelle Historie Militaire de la France* (Paris: Fayard 1998).

Stevenson, Roger, *Military Instructions for Officers detached in the Field containing a scheme for forming a corps of partisans illustrated with plans of the manoeuvres necessary in carrying on the petite guerre* (London: John Millan, Edward & Charles Dilly 1770).

Strickmann, Eva, *Clausewitz im Zeitalter der "neuen" Kriege* (Berlin: Galda 2008).

Sumida, Jon Tetsuro, *Decoding Clausewitz: A New Approach to On War* (Lawrence: UP of Kansas 2008).

Valentini, Georg Wilhelm Baron von, *Abhandlung über den kleinen Krieg und über den Gebrauch der leichten Truppen mit Rücksicht auf den französischen Krieg* (1st ed. 1799, 2nd ed. Berlin: J.B. Boicke 1820).

Waldtstätten, Baron von (ed.), *Erzherzog Karl – Ausgewählte militärische Schriften* (Berlin: Richard Wilhelmi 1882).

Weigley, Russell F., *The American Way of War* (New York: Macmillan 1973).

Atrocities in Theory and Practice

BEATRICE HEUSER

Classical counterinsurgency theory – written before the 19th century – has generally strongly opposed atrocities, as have theoreticians writing on how to conduct insurgencies. For a variety of reasons – ranging from pragmatic to religious or humanitarian – theoreticians of both groups have urged the lenient treatment of civilians associated with the enemy camp, although there is a marked pattern of exceptions, for example, where heretics or populations of cities refusing to surrender to besieging armies are concerned. And yet atrocities – defined here as acts of violence against the unarmed (non-combatants, or wounded or imprisoned enemy soldiers), or needlessly painful and/or humiliating treatment of enemy combatants, beyond any action needed to incapacitate or disarm them – occur frequently in small wars. Examples abound where these exhortations have been ignored, both by forces engaged in an insurgency and by forces trying to put down a rebellion. Why have so many atrocities been committed in war if so many arguments have been put forward against them? This is the basic puzzle for which the individual contributions to this special issue are seeking to find tentative answers, drawing on case studies.

Insurgencies, i.e. uprisings or rebellions against governments (whether these are constituted of an indigenous ruling group or a foreign occupation power) are a form of civil wars.[1] Like any civil war, they constitute a fight over the legitimate claim to exercise government. They are a special form, or at least they are civil wars defined in a special way. While the term 'civil war' suggests a certain balance between two forces claiming legitimacy for themselves while contesting the other party's legitimacy, an insurgency is a civil war in which one party (usually the government) is defined as exercising government 'legitimately' (in the case of the government, usually because it has traditionally enjoyed recognition by other governments), while the other party by this definition is a 'rebel' party, challenging the 'legitimate' rulers.

Insurgencies usually take the shape of irregular warfare, 'irregular' in many ways. One aspect of irregularity is that an insurgency rarely pits a rebel force against a similarly constituted government force in a battle between equals. Generally, the government side has the command of the 'regular' armed forces of the state, meaning they have trained, often professional, soldiers and are usually better equipped than the insurgents, who often rely on volunteers with little training. Such military training is important because it is supposed to bring about not only greater effectiveness and efficiency in fighting, but also because it usually revolves around the eternal paradox of modern state warfare, namely that it is supposed to be well-targeted and disciplined violence, not a random unleashing of it.

The characterisation of insurgencies as 'irregular warfare' is due in part to frequent cases in which restraints on violence are seen to be lacking, leading to what can broadly be termed atrocities, for which there are many historical examples. These can take place against enemy combatants, even though generally operational laws or rules that apply in 'regular' warfare prohibit their maltreatment, e.g. after they have surrendered, or when they are wounded.[2] Atrocities include needlessly painful and/or humiliating treatment of enemy combatants, beyond actions needed to incapacitate and disarm them. Atrocities also include the use of violence against civilian populations, i.e. non-combatants, especially where substantial numbers are affected. They include rape, mutilations, torture and killing in any way. One explanation is that a relative lack of, or inadequate, training sometimes accounts for atrocities committed by insurgents, as these are generally driven by emotion rather than fighting as disciplined military professionals. Nevertheless, there has also been a pattern of particularly cruel treatment of insurgents and their non-combatant sympathizers by the 'regular' forces of the government that is trying to repress them in counterinsurgency (COIN) operations. Professional discipline or the lack of it is not the only key to why atrocities occur; both insurgencies and COIN operations have been associated with atrocities throughout history.

Historical precedents may well show that the – in fact not at all 'new' – wars of the post-Cold War era are no more atrocious than many previous wars. Indeed, Melander, Öberg and Hall have argued that their impact on civilians has decreased compared with the wars of the Cold War. However, they do not deny that more recent wars have still produced unspeakable atrocities and suffering in intolerable numbers.[3] Thus, the subject clearly continues to be of major importance.

This article focuses on the subject of atrocities in insurgencies and COIN. The most important question about this subject is, 'why do they occur?' As will be demonstrated in this article, classical and neo-classical COIN theory, as well as most theory on how to mount and conduct an insurgency (insurgency theory), rule out atrocities as directly contravening the ultimate aim of establishing a stable peace. The practice, as will then be shown, has often diverged from theory. This article first surveys the evolution of COIN theory, then seeks patterns in the practice and finally addresses the question why theory and practice diverge, summarizing some of the answers provided by the case studies in this issue.

THE THEORY

COIN Literature

As we shall see below, COIN *practice* has mostly, throughout history, tended towards brutal repression. Classical COIN theory, however – literature dating from *before* the Age of Imperialism, Social Darwinism, and Racism – has with few exceptions cautioned strongly against the commitment of atrocities. This classical literature has been all but forgotten; most authors today seem to assume that literature on COIN has its roots only in the 19th century, a period marked by the rise of nationalism and <u>racism</u>.

Classical COIN literature

Let us start, then, with a survey of the classical COIN theory. Since late Antiquity, even before Christian arguments spread, some authors writing about how to conduct war underscored the need to show mercy to the defeated enemy's population, putting forward pragmatic reasons. Sextus Iulius Frontinus (c. 35–103 or 104 C.E.), for example, listed examples where the generous treatment of populations of conquered regions, including even rebel cities, had led to winning their allegiance.[4] We find the pagan Onosander, writing before 57 C.E., also emphasising the need to spare the lives of defeated populations and to avoid all cruelty, without citing any moral reasons.[5]

Christian authors tended to invoke divine pleasure to justify clemency, rather than to seek any more rational justification. As Christianity was for a century co-terminous with the late Roman Empire, St Augustine, as one of the early Church Fathers, could depict a world in which Rome's enemies were also Christ's enemies, and in which Rome (i.e. Christendom) was on the defensive against the onslaughts of heathens. In this context, collective self-defence against pagan aggressors could be legitimised. Any killing of non-combatants by contrast could not be legitimised, unless they were seen as worse than pagans, namely Christian heretics.

The religious wars of the 16th and 17th centuries – always civil wars, seen by both sides as a contest for legitimacy – were remarkable for their cruelty and large numbers of casualties. Even theoreticians took it for granted 'That a Prince may chastise by war, or otherwise, his heretical subjects', although some theoreticians began to urge 'some moderation, and not before he has convinced them not to be heretics by the word of God'.[6]

Quite analogously to the brutal repression of challenges to the Catholic Church and its monopoly on salvation, princes or other rulers in the Middle Ages and Early Modern times persecuted any challengers to their authority bloodily, so that COIN operations were almost invariably marked by atrocities. In line with the conviction that rebels were criminals, the French aristocrat Raymond de Beccarie de Pavie, Baron de Fourquevaux, explicitly noted in his instructions on warfare of 1548 that the normal laws against any ill treatment of the inhabitants 'of the country where the war is waged, either in body or goods' were suspended if these were 'declared rebels against the king'.[7]

Understandably, perhaps, the rulers' focus from Antiquity to the Present has lain on the danger emanating from those challenging their government and its authority, and rulers have consequently seen every reason not only to fight insurgents but to criminalise them in their propaganda and to punish them as criminals in order to deter any further insurgencies. From Early Modern times, however, several authors writing about how to occupy a country, prevent rebellions and respond to them if they occurred took a very different approach. Unlike princes, authors writing on the arts of war and statecraft (or, in today's parlance, governance), from Christine de Pizan (around 1400) to the present, could afford to take a more detached position, and to recognise that at least some insurgencies might be caused by poor governance: deficient rulers brought rebellions upon themselves. Niccolò Machiavelli (1469–1527) explained a government's fear of its subjects as having its cause in the hatred which these subjects must feel for the government on account of poor governance.[8] And while Livy (59 B.C.E.–17 C.E.) had cavalierly suggested that 'subject peoples should be generously treated or wiped out', Machiavelli could not bring himself to support the latter option.[9] Following his countryman Machiavelli's example, Giacomo di Porcia (1462–1538) therefore urged his commander to issue instructions against maltreating civilian populations in conquered areas who might otherwise rise up against their oppressors.[10]

Towards the end of the 16th century, the British cleric Matthew Sutcliffe (c.1550–1629) drew from the Ancients the insight that 'a subdued country is kept by the same means that subdued it; that is by fortitude, industry, *and justice* [my emphasis]'. Therefore, his prescription for the occupation of a country was, 'after the victory, . . . [t]o keep our conquest, there are two principal means which are necessary; force and justice. For neither can those who are rebellious, and desirous of change [*innovation*], be repressed without force, nor can the peaceful be defended, or contented without justice . . . And it is hard to keep discontent men in subjection for long by force'.[11] Again, Sutcliffe reiterated his recognition that poor governance would cause rebellion:

> For the rest, if the governors of newly conquered countries should be careful and watchful, and trust no man without reason, and use equality in taxation, and enact good justice against thieves [*ravevours*], bribe-takers, and rebels, they need not fear rebellion; if they do not, all force that may be used will not serve to keep them in subjection for long . . . For no people can like a government for long, when their property is spoiled and they are vexed, injured, and to say all in one word, pillaged, and tyrannised.[12]

This recognition would echo through the centuries[13] and was periodically seized upon by political philosophers, from Jean de Gerson, Hugo Grotius and Algernon Sidney to John Locke, and by political leaders of rebel camps to justify rebellions.[14] The most commonly contemplated scenarios were thus, in times of peace, uprisings against lords and governments ruling from afar and perceived as foreigners; or, in the aftermath of a war, uprisings of conquered peoples against their new rulers; or, within warfare, the refusal of a town or city to surrender to an army that lay siege to it.

Sieges

COIN has its roots in past theories and practices of siege warfare, which were always seen as a special case of rebellions and attempts to quell them. Since Roman times, sieges have implicitly represented a dispute about legitimate rule. Any town that was besieged, rather than granting the besieging forces entry, by the very act of defying these (and not declaring itself an 'open city') stood up against the authority for which the besieging forces were fighting, and thus fell within the category of rebels. The context was always assumed to be a struggle between two sides for the legitimate rule over the town under siege (and usually other areas), where war amounted to the settlement of a lawsuit. War, in this case seen in the medieval tradition of a divine ordeal, could only establish the justice of one side, and, thus, the other and all who supported it were branded as rebels and criminals. In general, it was held that a besieged town that had surrendered must not be ransacked and its citizens must be treated with respect. If a castle, town or city refused to yield in a siege, however, and was then taken by force, it was usual to maltreat the obstinate citizenry horribly; even strategists were not always opposed to this. Porcia, whom we have encountered above as urging good governance for conquered areas, in 1530 advocated cutting off the hands of the townsfolk who had resisted a siege in order to coerce other towns into a swifter surrender, even though he also urged his reader to hold out in any siege. Moreover, he counselled increasing the besieging army's fervour by promising them that they could pillage to their heart's delight.[15] Such atrocious practices, known from Africa in the 19th and 20th centuries (see below, and see Kieran Mitton's article in this issue), have horrifying European antecedents.

In contrast, Bernardino Rocca from Placentia, writing half a century later, cautioned against losing the fruits of one's victory, which would be the case if the soldiers were allowed to commit atrocities.[16] The Englishman Robert Barret (1598) tried to find a compromise. In his text, he has his Captain explain to the Gentleman that in taking a 'fort, city or town', the military commander

> should pursue the victory even until the enemy wholly yields, and rendered, and grants license to fall to sack and spoil: after which he shall deport himself in neither a cruel nor covetous manner, as a number of bad and graceless fellows do, without respect for God or man, leaving no kind of ravening cruelty uncommitted, with brutal ravishment of both women and maidens, and with merciless murder of poor innocents: instead in such cases he shall show himself to be favourable and merciful to the humble vanquished, making arrangements to defend them, especially simple women and maidens: for God, no doubt, will be well pleased in so doing.[17]

Barret's countryman and contemporary Sutcliffe recalled Caesar's command to his soldiers to spare Rome even after Rome had sided with his defeated enemy Pompey in the Civil Wars.[18] In the French wars of religion, the conciliatory King Henri IV also tried to protect civilians from reprisals. Sutcliffe found this behaviour befitting Christians. He noted that even the Spanish, whom he heartily detested, in their attempts to quell the Dutch rebellion, had eventually espoused the norm that 'Women, children, and the elderly, by the order of war now observed in the Spanish camp, are exempt from the soldiers' fury in the sack of towns'.[19] Consequently, Sutcliffe drew up legislation to prevent, or else severely punish, any transgressions against the local population.[20] He added:

It is inhumane and harsh to massacre those that surrender, and throw down their weapons, confessing themselves to be vanquished, and flying to our mercy ... The Spanish in the beginning of their wars in the Low Countries cruelly killed as many as they took. But when they saw themselves to be dealt with in a similar way, they repented, and perceived such savage cruelty as contrary to the nature of fair wars.[21]

He did not mention what was to be done with the inhabitants of towns that had been taken by force. Barret's and Sutcliffe's contemporary, Alberico Gentili, in his *Three Books on the Laws of War* of 1598, likewise cautioned against the use of force against non-combatants – with the exception of a town taken by force, where the prevailing army would not be restrained.[22]

Another contemporary, the Spanish soldier and diplomat Don Bernardino de Mendoza, had witnessed in person the Dutch Rebellion and the Duke of Alba's bloody COIN campaigns. At the end of his life, in 1595, Mendoza wrote in his handbook on war:

While the authority of an empire and government is strengthened by the use of force, it is very dangerous to employ force against your own subjects, unless to all human reason it seems certain that you shall eventually punish them, lest you arm the lion with claws ...

He thought these were 'sufficient words to move a prince not to cover his sword in the blood of his vassals'.

Instead, he should put it off it with secret means of negotiation, of the sort that may be offered to reduce some of the principal heads of the insurrection, by treaties of grace and pardon ... The barbarians hold it to ... be the better part of a victory to restrain just fury and the anger of men from shedding blood among those that had previously loved each other ... This is a passion which can hardly be cooled, except if one considers that the taste for revenge only lasts a few days, and the joy of piety is eternal. Therefore let justice be accompanied by clemency, so that it is not cruelty, and let clemency be accompanied by justice, so that it will not be held in contempt. In the insurrections of cities and provinces, or mutinies of soldiers, [the prince] is obliged to punish those who have initiated the movement, being the authors thereof, and to pardon the rest, as it is not possible to punish a multitude.[23]

Mendoza's words stand in stark contrast with the bloody practice he had witnessed under the Duke of Alba, which had corroborated the *leyenda negra*, the Black Legend of the maltreatment of Spanish subjects in colonies by the Spaniards, originating with accounts of Spanish atrocities towards native populations in America.[24]

The same line of cautioning against atrocities can be found in the second half of the 17th century, in the writings of the Italian Marques Annibale Porroni, a mercenary who served as a major general in Poland. In his *Universal Treatise on the*

Modern Military inspired by the style of Machiavelli's *Art of War*, he urged against maltreating the population even after a successful siege.[25] At much the same time, the Frenchman Paul Hay du Chastelet gave the following advice to his king, Louis XIV, who for his many wars of conquest was dubbed 'the King of War':

> the conqueror has to ... caress the inhabitants [of conquered towns], giving them the hope that they will be happy under the new domination, and he has to deprive them of the memory of their previous situation, and he must give them reason to think that they will be happier under their submission than they were before; that they will be ruled according to their old customs, and that their privileges will be increased, and that they will suffer no reduction of their status [*diminuition*].

In musing on how to deal with occupied territories, he noted:

> It is not enough that a king, after having made a conquest, or having imposed peace on his defeated enemies, should think of repairing the evils which war can have caused in his kingdom. Beyond that, he has to turn his attentions diligently to the peoples he has conquered. For in becoming his new subjects, they have become his new children. It is thus necessary that he should make them forget the domination under which they have been born, and under which they had become accustomed to live. At the same time he has to dispel the pain of their defeat and that which they might have in obeying to the laws of the conqueror. These newly conquered peoples have to be happy about the change of masters, and they must believe to have reason to fear the return of the previous government. In order to achieve this positive result, the conqueror must govern his new subjects with all possible mildness ...

By which he meant, through measures such as tax relief, investments in the local infrastructure and public buildings, respect for their customs. Generally, the occupying forces 'must not commit acts of violence or injury against any private person. Those guilty of such lawlessness must be chastised in such a form that the injury is compensated and the act of violence repaired'.

Only if this kind approach did not bear fruit, and if the people continued to resist the new authority, did Hay du Chastelet advise the use of 'force and severity':

> one has to disarm them, change their laws, take away from them what is most precious and holy to them, deport them and colonise them, dispossess them of their lands, marry their daughters to the soldiers of the conqueror, forbid them all commerce, take away all their artisans, destroy their public buildings, keep their hostages, impose extraordinary taxes and keep armies in their midst in order to suppress them with more authority.[26]

Side by side with outlining a policy of winning 'hearts and minds', as this would be termed later, we thus find the prescriptions for punishment and humiliation, even ethnic cleansing, and, worse than plunder, dispossession of their property, in case the population proved unresponsive to a policy of kindness.

The Age of Enlightenment gave a boost to more humane views, expressed in the writings of the Spanish aristocrat Santa Cruz in the 1730s:

> If one is victorious in combat, one hast to spare the blood of the vanquished. They may be rebels now, but may soon become most loyal subjects. Only those who make a show of obstinacy should be punished severely, but not if the desperation is so great that making a bloody example of them will not deter [them] but make [them crave] revenge. The more aristocrats that were engaged in the insurrection, the less beneficial strict and gruesome punishment will be. Only if the rabble is rioting does violence often help more than kindliness.[27]

Violence, for Santa Cruz de Marcenado, was a last resort, to be shunned as long as there were still any other options. His long deliberations on the subject of how to suppress insurgencies chiefly turned on the need to separate the rebel leaders from the body of the population, and to win the latter over through acts of kindness, good governance and just rule.

After the Napoleonic Wars, the French general Le Mière de Corvey provided a late example of classical COIN theory, trying to base a new doctrine for small wars on the lessons drawn from the anti-French uprisings. He already fully articulated the need to win over the hearts and minds of the populations of occupied territories, writing that 'if you want to preserve your conquest, you must treat the defeated with clemency [*douceur*], this is the only way to win them over to you [*se les attacher*]'.[28] So far the classical COIN literature.

Imperialist and racist COIN literature

Notwithstanding such late examples of classical COIN theory as that of Le Mière de Corvey, here, as in major war, the French revolutionary and Napoleonic Wars generally created a watershed in thinking about war.[29] Classical COIN theory gave way to the more commonly known imperialist–racist COIN theory. Its authors, whose thinking was conditioned by the practices that they had witnessed, practices which they in turn had influenced, had a much less benign attitude towards local populations than their classical predecessors. In 1850, the French general Thomas Robert Bugeaud de la Piconnerie, Duke of Isly, who had witnessed the Napoleonic repression of the Spanish Guerrilla (i.e. small war) and transposed his experience to French colonial practice in the 1830s and 1840s, wrote that, to defeat the mountain peoples in Algeria,

> one has to attack their interests. You can only succeed ... by destroying their villages, cutting down their fruit trees, burning or uprooting their harvests, emptying their silos, searching their ravines, rocks and caves to seize the women, children, old people, herds and movable goods. Only thus can one bring the proud mountain people to capitulate.[30]

Earlier, he had pretended to be humane governor, when in 1845 he addressed the Arabs and Kabyles, saying:

The king, master over all of us, wants his Arab and Kabyl subjects to be as well governed and happy as the French ... In order to obey his orders and the inspiration of his fatherly heart, I exhort you as follows. The foremost way to repair the damages of war and to be happy is to remain faithful to the promise you have given us, and for which we have stopped our ... battalions. You must honestly accept God's command who has wanted us to come and govern this country. You know of the evils that have befallen the tribes that have revolted against us and against God's will.[31]

Disdain for the civilian populations in colonies seems to have been the norm in the 19th century among military strategists, and Bugeaud's countryman Marshal Lyautey was in the minority when stressing the idealistic form of European colonialism, the *mission civilisatrice* or Kipling's 'White Man's Burden', aiming to bring the local populations the benefit of European civilisation rather than oppression and tyranny.[32] Lyautey's colleagues, by contrast, and the governor with whom he had to work closely, thought that the 'oeuvre de la civilisation' was nicely complemented by a little punitive bombing of rebels here and there.[33] Equally in Britain, Charles Edward Callwell, famous for having written the earliest handbook in English on how to suppress uprisings, fully approved of Bugeaud's and the French attitude to insurgents in Algeria in the first half of the 19th century, and cited them as a model for what he himself prescribed.[34]

The rediscovery of hearts-and-minds campaigns

The importance of 'winning the hearts and minds' of insurgents only re-emerged in what might be termed neo-classical COIN literature after the Second World War, even though its authors lived in complete ignorance of the pre-Napoleonic antecedents of their views. The term itself is generally attributed to the British high commissioner Sir Gerald Templer, who successfully steered the British COIN operation in Malaya through its crucial phase and towards its ultimate success.[35] It resonated in the context of the Vietnam War, when official American rhetoric was so full of it that cynics and flippant military parlance turned it into 'WHAM-ing' the local populations. The rhetoric, however, was unmatched in quality or quantity by practice, a key criticism levelled at the US even in the 1960s.[36] The need to 'WHAM' became a consensual creed in expert debates about COIN in the early 2000s, enshrined in the famous American ('Petraeus') Field Manual 3–24 of December 2006,[37] although, again, practice would not always match expectations.

Similar lessons were drawn from the Vietnam War by its perhaps most famous war correspondent Moshe Dayan, as Eitan Shamir shows in his contribution to this special issue. Dayan, at once theorist and practitioner, represents the strong strand in the Israeli counterinsurgency tradition which tried to win over non-Jewish minorities within Israeli-controlled territories – the Druse in the Golan Heights, the Arabs in Gaza – by improving their lives and giving them a stake in the economic and political settlement.

Insurgency Literature

Literature on how to stage insurgencies, which began to appear much later than literature on how to fight against them, shares with both classical and most recent COIN literature the goal of winning the support of civilian populations through all possible ways other than terror-inspiring atrocities. There is a logic to this, as most of the literature was written by ideologically driven men who saw themselves as fighting on behalf of oppressed peoples, and it is in keeping with this perception that they would call for good treatment of these populations. In general, the assumption was that if the civilians were not on the side of the insurgents, or were neutral, this was through ignorance and should be addressed by means of persuasion and education, not by terror or brutal repression.

Several of the earliest authors dealing with the organisation and execution of insurgencies did not dwell in any way on the treatment of civilians. This was the case probably either because they shared this assumption about the tacit support of their own populations against foreign occupants (as in the case of the Prussian Clausewitz[38] or the Pole Chrzanowski[39]). Or else, very importantly, the treatment of civilians was not addressed because the enemy was a foreign occupation force on one's own soil, and enemy civilians were on the whole nowhere near the theatre of operations, and one's own civilians were obviously allies. Once the transition of the concept of 'small war' from special force operation to insurgency had been made, few saw the need to spell out but presumably took for granted what the Latin American Communist leader Ernesto 'Che' Guevara articulated in 1959:

> The *guerrillero* ... is the freedom fighter *par excellence*; he is the elected of the people, its fighting vanguard in the fight for liberation. *[G]uerrilla* warfare is the warfare of the entire people against dominant oppression. The *guerrilla* is its armed vanguard; the army is made up of all the inhabitants of a region or country. This is the reason of his force, of his triumph; ... the base and the substratum of the *guerrilla* lies in the people.[40]

It stands to reason, by such a definition, that the *guerrilleros* must not treat their own people as enemies.

Most writers on how to wage a small war, going back even to the 18th century, took for granted the view that one should be on as good terms as possible with the local population. In the American War of Independence, the German mercenary Johannes von Ewald, fighting on the British Hanoverian side, emphasised the need for discipline among partisan forces – here the term still used in the sense of special forces fighting alongside regular forces, under the authority of a state.[41] One can sense from his writings that Ewald himself was already experiencing the transition of small war operations from special force operations to people's wars over the legitimacy of a government; for in this war, the American rebels employed not only regular forces, but also irregulars, and fought with growing support from the civilian settlers. Ewald's emphasis on the need to enforce the strictest discipline on the partisans fighting for British rule likely reflects his realisation that the Royal Army

155

no longer had the unconditional support of the local population. Ewald admonished the 'partisan', a term still employed only for the leader of the *parthey* or special forces unit, to ensure that

> Not the slightest infringement upon discipline, orderliness and service must be tolerated, especially not in the beginning ... Above all one cannot deal harshly enough with those villains who mercilessly torment the peasants who are innocent of the war. The best thing to do is to chase such rabble away, since those who once stooped to plundering can never be trusted again ... Do not believe that you can gain the love of a soldier through an unpermissible kindness and indulgence at the expense of the poor peasant and by a policy contrary to all human nature.[42]

The Prussian general von Valentini, writing a decade later, shared this preoccupation with the benign treatment of the civilian population, even though he was more concerned with discipline than with winning hearts and minds.[43]

In the Spanish Guerrilla (1808–12), which famously redefined the Spanish term for 'small war' as 'insurgency', the rebels fighting against the French were fully inspired by the conviction that they were defending Spain and, thus, an imagined ideal community of ideal Spaniards (Catholic, royalist and full of solidarity for one another). The reality was that many local scores were settled in this context.[44] Therefore, even during the Guerrilla itself, the need was seen to issue rulebooks for the *guerrilla* fighters (Spanish: *guerrilleros* or *somatenes*), which included the strong admonition to differentiate between the treatment of enemy soldiers on the one hand and the population on the other, whether or not they sided with the Guerrilla against France. The need for disciplined behaviour in the villages and other settlements was underscored.[45] In his diary of the events in Spain of 1811–12, Don Francisco de Copons y Navia also stressed how the heroic actions of the fighters (here called *somatenes*) had to be seen in the context of the severe punishment that had to be inflicted on the disobedient and especially those who had committed 'excesses'.[46]

Like Valentini before him, the leader of the Italian uprising against foreign rule and architect of Italian unity Guiseppe Garibaldi, in his *Decalogo tattico* of 1848, was more concerned with the morale of the guerrilla fighters themselves than with the support – tacit or active – of the population.[47] Only in 1870 did he emphasise that the partisans must make themselves respected and loved by the population, but he did not elaborate on the consequences that this might have for the sort of treatment that might be meted out to them.[48] In the Arab rebellions which T. E. Lawrence helped organise in the First World War, again, the main body of Turkish civilians were far away, and tribes tended to stick together. So the problem of how to deal with hostile non-combatants was not crucial.

This applied also to China after the Japanese invasion of China in 1937. Seeing his Communist Party as one fighting in the interest of his people, Mao Tse-tung issued a code for practice for the Chinese war of liberation against the Japanese that year, which included 'Three Rules and Eight Remarks', running as follows:

Rules:

1. All actions are subject to command.
2. Do not steal from people.
3. Be neither selfish nor unjust.

Remarks:

1. Replace the door [often used as a bed] when you leave the house.
2. Roll up the bedding on which you have slept.
3. Be courteous.
4. Be honest in your transactions.
5. Return what you borrow.
6. Replace what you break.
7. Do not bathe in the presence of women.
8. Do not without authority search the [bags] of those you arrest.[49]

Reading this, it is difficult to predict that the author of these Rules and Remarks would one day become one of the greatest mass murderers that the world has seen. Crucially, however, when Mao was writing this, the non-combatants were almost entirely on the side of the Chinese resistance against the Japanese imperial army, and Mao in his writing did not have in mind a civil war in which there was burning hatred against the part of the population that sided with the enemy.

Emulating Mao, Ho Chi Minh in Vietnam issued 'Twelve Recommendations' on 5 April 1948 which were proclaimed by his generalissimo Vo Nguyen Giap:

Six Forbiddances:

1. Not to do what is likely to damage the land and crops and spoil the houses and belongings of the people.
2. Not to insist on buying or borrowing what the people are not willing to sell or lend.
3. Not to bring hens into mountainous people's houses.
4. Never to break our word.
5. Not to give offence to people's faith and customs (such as to lie down before the altar, to raise feet over the hearth, to play music in the house, etc.)
6. Not to do or speak what is likely to make people believe that we hold them in contempt.

Six Permissibles:

1. To help the people in their daily work (harvesting, fetching firewood, carrying wood, sewing, etc.).
2. Whenever possible to buy commodities for those who live far from markets (knife, salt, needle, thread, pen, paper, etc.).
3. In spare time to tell amusing, simple, and short stories useful to the Resistance, but not to betray secrets.

4. To teach the people the national script and elementary hygiene.
5. To study the customs of each region so as to be acquainted with them in order to create an atmosphere of sympathy, then gradually to explain to the people to abate their superstitions.
6. To show to the people that you are correct, diligent, and disciplined.[50]

Again, these rules originated in the Indochina War against a foreign occupation force (France), rather than the Vietnam War, which was to a much greater extent a civil war. The practice of North Vietnamese treatment of the South Vietnamese in the subsequent conflict was very different. While South Vietnamese government publications on this subject must be treated with some caution, as they were not too careful to avoid hyperbole and unverified accusations, the pattern of small-number atrocities against children, women and unborn foetuses as illustrated here has much in common with those we find in the case studies of Mitton, Reis and Oliveira in this issue.[51]

Twenty years later, Che Guevara wrote in terms of 'annihilating' the enemy, but this seems to have been more a figure of speech than 'annihilation' in a Ludendorffian, national–socialist sense. To Che Guevara, the 'enemy' clearly excluded civilian populations. Indeed, he even advocated a lenient treatment of enemy combatants, once they had been defeated: 'A fundamental part of guerrilla tactics', he wrote, 'is the treatment accorded to the people of the zone'. Furthermore:

> Even the treatment accorded to the enemy is important; the norm to be followed should be an absolute inflexibility at the time of attack, an absolute inflexibility toward all the despicable elements that resort to informing and assassinating, and clemency as absolute as possible toward the enemy soldiers who go into the fight performing or believing that they perform a military duty. It is a good policy, so long as there are no considerable bases of operations and invulnerable places, to take no prisoners. Survivors *ought to be set free*. [my emphasis]

And he clearly took his Hippocratic oath taken as a trained physician seriously when he added,

> The wounded should be cared for with all possible resources at the time of action. Conduct toward the civil population ought to be regulated by a large respect for all the rules and traditions of the people of the zone, in order to demonstrate effectively, with deeds, the moral superiority of the guerrilla fighter over the oppressing soldier. Except in special situations, there ought to be no execution of justice without giving the criminal an opportunity to clear himself.[52]

While Che entirely bungled his own operation to introduce a Communist revolution to Bolivia, as well as Argentina and Peru,[53] in contrast to Mao or Giap, there are few recorded cases where he stands accused of deviating in his practice from his beliefs (one of them his terror regime in the La Cabaña Prison in Havana).

THE PRACTICE: ATROCITIES, PLANNED AND SPONTANEOUS

To sum up, most literature on how to organise and wage insurgent warfare, just as the classical and neo-classical COIN literature on how to counter it and most of the most recent works, strongly advise against permitting atrocities, let alone planning them, both for purely practical reasons and on moral grounds. Yet in practice, many insurgency leaders throughout history have used atrocities to terrorise populations into co-operating. The same holds true to those suppressing insurgencies, who more often than not have either permitted or even planned atrocities that have been recorded in history, and there are likely to be many more unrecorded instances.

There is not space here to cover all cases of atrocities committed in insurgencies or COIN operations throughout history, nor could any quantitative statement be made authoritatively.[54] It lies in the very nature of massacres that they are often forgotten in history precisely because none of the victims live to tell the tale, and given the general code of practice, because the perpetrators were often none too proud of what they had done, or feared the reaction of some public opinion or higher moral authority. One should just think in this context of how many mass graves of the Second World War, or of various earlier or later phases of Soviet rule, were only discovered by accident, most infamously that of the Katyn massacre. The following is thus merely an attempt to sketch some patterns. No pretence is made that a quantitative judgement can be arrived at: we cannot say whether *most* or just *some* insurgencies (or COIN operations) were blackened by atrocities, i.e. whether the known atrocities were exceptions or the rule. All that we can say is that atrocities *have* occurred time and again. When recorded by historians or other sources, it is in many cases difficult to tell whether these authors noted them as outstanding exceptions, or (as is the case in Grimmelshausen's *Simplicissimus*[55]) as the norm, albeit one they themselves might condemn.

Planned Atrocities against Rebels

Clearly, many atrocities were actually planned well in advance on the side of COIN forces, and thus clearly and deliberately part with classical COIN theory and its prescriptions. In Antiquity, little restraint seems to have been practised in dealing with unarmed populations that were identified with the enemy, including children, the old and infirm, and the women who did not participate in the fighting in most societies of that age. Written sources like the *Iliad*, the Hebrew *Bible*,[56] and later Greek and Roman historiography abound with examples of non-combatant populations being killed or abducted as slaves. On the whole, it was men who were massacred, while women and children were taken as spoils.[57] Hans van Wees points out that the mutilation of the corpses of soldiers was regarded as a horrendous infringement of the laws of war in classical Greek warfare, but again, the *Iliad* gives us several examples of just such actions during a siege.[58] The Romans generally punished rebellions of colonised peoples with sufficient cruelty to prompt some of them to commit suicide rather than to surrender; this was the case of the peoples of Numantia on the Iberian peninsula in 133 B.C.E., and Masada in Palestine in 73 C.E.

Roman massacres of non-combatants occurred in such great numbers that it is difficult to pick out one example over another. Ceasar, after the fall of Avaricum (Bourges) in Gaul which had defied him, had an entire population of almost 40,000 men, women, and children killed. On the one hand, Caesar permitted this atrocity in revenge for a Celtic atrocity committed by the Gauls at Cenabum (Orléans); on the other hand, he used this instance to put psychological pressure on the citizens of Gergovia, who also had joined the insurgency of Vercingetorix, to surrender to the Romans.[59] Another particularly infamous example of Roman brutality was that of the repression of the Jewish rebellions of 66–73 C.E. (the First Roman–Jewish War) and 132–135 C.E. (the Bar Kokhba Revolt), seen as the true beginning of the scattering of the Jews throughout the diaspora.

By contrast, in Buddhism, there have always been qualms about killing, and in the Mediterranean region with the spread of Christianity, qualms developed about killing fellow Christians. Soon thereafter, however, the fight between 'orthodoxy' and 'heresy' began to tear the late West Roman and the East Roman Empires apart and arguably drove populations branded as heterodox, or heretical, into the arms of the occasionally more tolerant Muslim conquerors. The rulers of Constantinople had already resorted not only to persecution, but also to ethnic cleansing or population displacement, most famously of adherents of dualist creeds, the Paulicians or Bogomils, who were evacuated from the eastern end of the Empire to the Balkans in order to forestall collaboration with the Empire's Persian enemies. In western Christendom, the persecutions of the Cathars (possibly related to the Bogomils) in the Albigensian Crusade or the early 13th century mixed religious war with political motivations, the outcome of the bloody repression of the Cathars was that their leaders' lands fell to the French crown. Then, from around 1400, came the first wars between Catholics and early Protestants: the Hussites in Bohemia, and the Lollards in England. These were prosecuted by Catholic princes and other Catholic polities in ways which suspended normal conventions of respect for the adversary, his combatants and non-combatants alike, substituting greater cruelty, as was normally reserved for 'barbarian' cultures.[60] All these examples show a pattern of *planned* atrocities.

In a non-religious context, the Great Invasions, as the period of the mostly violent migrations from eastern Europe and central Asia to western Europe and into the Roman Empire is called in French, were often accompanied by mass slaughter of local populations. In those centuries, measures to quell rebellions (or what was presented as such) were brutal, in keeping with more general patterns of the time. Gregory of Tours' history of the internecine feuds of the Frankish rulers is full of devastating punishments wrought on the populations under one ruler challenged by another, usually brothers, cousins, or other members of the same family through intermarriage.[61]

If rebels refused to give up their ways, annihilation of unarmed subject populations of the rebels, or of their combatant supporters – by direct killing or indirect *planned* scorched earth tactics – continued to be part of a legitimised strategy, famously in Charlemagne's massacres of the Saxons to the east of his expanding realm, but it helped the public presentation of his case that the Saxons had resisted Christianisation and were, thus, pagans.[62] This pattern is still found in the High and Late Middle Ages, where the customary punishment for disobedience among subject peoples was to follow the rebels to their own settlements, and there to 'drain the swamp', to use an early 21st century metaphor, by killing all and sundry. Medieval princes thus punished populations for not accepting their own claim to over-lordship. At first, the Normans met with relatively little opposition in England, once the main host under King Harold had been soundly defeated near Hastings. But only 3 years on, there were uprisings in the north of England, which were put down by the Normans in the infamous – and well-planned – Harrowing of the North, in which it is estimated that 150,000 people were killed or died as the result of famines caused by Norman scorched earth policies.[63] (Despite this repression, the struggle continued in the Fens around Ely, with the invasion of the Danish king Sweyn II Estrithson, nephew of Canute the Great, himself born in England. Sweyn tried to challenge the rule of his Norman cousins and was joined by local rebels, including Hereward the Wake, until the fall of Ely in 1071.)

We find this pattern particularly prominently in the English kings's reactions to the Welsh uprisings. Before the Norman Conquest, the Welsh had struggled to cast off Anglo-Saxon over-lordship, and 3 years before the Norman Conquest, Harold Godwinson, the future competitor for the crown of William, had sent a punitive expedition to Wales, 'slaying every male who could be found, even down to pitiful boys, ... thus [pacifying] the province [of Snowdonia] at the mouth of the sword'.[64] The Normans continued to treat the Welsh in the same fashion. In 1094, William II 'Rufus', son of the Conqueror, dealt harshly with the Welsh insurgents against Norman rule, led by Gruffydd ap Cynan. Rufus 'intended to abolish and utterly destroy all of the people until there should be alive not so much as a dog'.[65] In 1116, a Welsh force loyal to Rufus' brother and successor Henry I promised in their fight against the rebel forces of Gruffydd ap Rhys 'not to spare the sword against man or woman, boy or girl, and whomsoever they caught they were not to let him go without killing him, or handing him, or cutting off his members'.[66] More examples of such planned massacres can be found, especially in the conflicts between the rulers of England and their British neighbours, and on both sides, and Matthew Strickland has underscored the similarities to patterns of 19th and 20th century colonial warfare: on the one hand, the disdain of the kings and armies of England – who were used to fighting quite differently against their 'peers' in France – for the Welsh, Scots and indeed Irish, who were after all also Catholics, and on the other the almost nationalist hatred of the latter for the English (whether their rulers were Anglo-Saxon, Norman, Plantagenet or Tudor). But examples can be found of English treating English with similar disdain for the church-imposed rules of sparing civilians. Again, typically in a civil war (or insurgency) scenario, the Earl of Chester, fighting against King Stephen in the 12th century war over the succession to the crown of England:

began ... everywhere to rage cruelly with plunder and arson, violence and the sword, sometimes against his opponents, sometimes even against his own side ... [H]e passed rapidly from one region to another with his unbridled army and by his ravages turned everything into a desert and bare fields.[67]

While it may be true, as Matthew Strickland has argued, that the English by and large respected rules of war in their dealings with French princes and their regular armies, which they disregarded in the British Isles, they similarly inflicted plunder, arson and scorched earth tactics – always a premeditated measure – on the peasantry of France during the Hundred Years' War.[68] Again, the peasants, if they refused to submit to the Plantagenet kings, could be defined as rebels.

Planned massacres to punish rebels and insurgents were also ordered by the rulers of France, when departures from the normally observed rules of 'intracultural' war could be suspended by the fight against a rebellious heretical 'subculture', to use Hans-Henning Kortüm's classifications (see below). The Albigensian Crusade mentioned above also has to be mentioned in this context. At the height of the wars of religion which tore continental Europe apart in Early Modern times, the St Bartholomew's massacre in France in 1572 was a case of a clearly planned atrocity, backed by the King Charles IX himself.[69] This was neither the first nor the last planned massacre of Huguenots, while the latter can be charged at least with spontaneously committed atrocities, or planned and targeted assassinations of individual Catholic leaders. It is not accidental that the St Bartholomew's massacre occurred during the years when the third Duke of Alba governed the Spanish Netherlands (1567–73), with bloody repression of the Protestants who were challenging Spanish rule.[70]

In the following century, perhaps the most famous large-scale atrocities were those committed after the siege and fall of the city of Magdeburg in 1631 during the Thirty Years' War, which stood in the tradition of Medieval atrocities committed to punish a city that had refused to surrender. This was a case of poor leadership, fitting the pattern identified by David Benest in this special issue, when the Habsburg general Tilly failed to control the troops present there (the Pappenheimer, called after their commanding officer, and the Walloons), all of them professional soldiers. Total figures of dead are uncertain, but estimates vary between 20,000 and 25,000, with many more victims raped, injured and suffering through loss of property or bereavement. Contemporary sources testified that neither children nor pregnant women were spared, and that the atrocities caused revulsion even among some other units under Tilly's control.[71] This is much the same pattern which Kieran Mitton found in his study of Sierra Leone, further on in this special issue, and similar atrocities were committed in the Yugoslav Wars in the 1990s; they are thus not specific to Africa. The civil wars on the British Isles in the mid-17th century also fitted the pattern that where ideologies motivated parts of the same population to fight each other, atrocities happened easily and spontaneously. [72]

Premeditated atrocities, by contrast, can be found again in the context of the wars between the Protestant rulers of the United Kingdom of Scotland and England, and the Catholic Scots and Irish who defied them, at the very end of the 17th century and during the Jacobite Rebellions in the early 18th century. The Age of Revolutions brought insurgencies in America, and throughout Europe, first against British colonial rule (the American War of Independence, 1775–83), then against the French Revolution (in the Vendée uprisings 1793–96), then against French occupation under Napoleon (in Spain and Russia, 1808–12); and, at the end of this spiral of reciprocally inflicted violence, the 'White Terror' of 1815, in which French Catholics wrought their revenge against local adversaries who had flourished under Napoleon.[73] They all had in common that in the treatment of the rebels and their dependents common rules of conduct, *ius in bello*, were suspended: normal 'rituals of surrender' were not followed from the moment that the defeated enemy were seen as rebels.[74] Instead, the soldiers were humiliated, and in the case of the Jacobite Rebellions they were hounded to death even after fleeing the battlefield, wounded Jacobites were bayoneted and many civilians suspected of Jacobite sympathies were killed. Remaining Jacobites were rounded up and imprisoned; those who survived the harsh conditions of imprisonment were deported.[75]

The record of planned atrocities was greatly added to in the 19th and 20th centuries. We have already quoted Bugeaud's prescriptions that were put into practice in Algeria in the 1830s and 1840s. These are not exceptional among the colonial policies of the European colonial powers in the 19th and early 20th centuries, as Jacques Frémeaux has ably shown.[76] In his article in this issue, he indicates that even in the 1950s, there was some deliberate use of terror against the civilian population in Algeria so as to deter them from siding with the insurgents.

The history of rebellions of the Christian peoples under Ottoman rule in Europe throughout the 19th and early 20th centuries is packed with local atrocities against individual communities committed by both sides. It culminated, however, in the large-scale Turkish genocide of the Armenians, who the Turks claimed (and might even to some degree have convinced themselves) were rebels against Ottoman rule and a Russian fifth column. Whether this was objectively the case or not is irrelevant; the link with a rebellion was used to justify the maltreatment of the Armenians.[77]

In African colonies, European powers also committed premeditated atrocities. Similar to Muslim practices of punishment for theft, the (Christian) Belgian administrators of King Leopold's Congo inflicted large-scale mutilations (especially the severing of hands) of indigenous labourers who had failed to fulfil their work quota.[78] Again, the examples quoted by Kieran Mitton are not specific to indigenous African practices, Europeans followed them as well. In other cases of atrocities committed in Africa, the cutting off of hands has been linked specifically with Islamic Sharia law. As recently as 2002, it was applied by Sudanese Muslim administrators in South Sudan to terrorise the local black and non-Muslim population into submission to the Arab/Muslim-dominated central government. Other examples of such atrocities include those in Liberia under Charles Taylors' regime of terror, where they cannot be accounted for in terms of a Muslim regime. And as we have seen, the idea itself can be found in Christian Europe in the 16th century, in the above-mentioned writings of Count Porcia.

Another category of planned atrocity was that of the engineered famine and resultant mass starvation, or related death by exposure. Engineered famines are of course really a modern version of the scorched earth tactics that go back to Antiquity. Along with direct killing, engineered famines were applied at the beginning of the 20th century by the Germans against the Hereros and Nama in their African colonies, and subsequently by the Turks against the Armenians.[79] Mass starvation was also the measure used by Stalin to subdue Ukrainian nationalism in 1932–33, which has become known as the *Holodomor*, killing by hunger. Mass starvation was furthermore used by the Nigerian government against the Igbo to counter their attempts to gain independence in 1967–70 by setting up the Republic of Biafra.

Atrocities were committed by both sides in the Spanish Civil War (1936–39). While there seems to be consensus today among historians that the victorious Fascists under General Franco with their 'White Terror' caused the deaths by execution, aerial bombing or otherwise of several times as many as the Republican forces with their 'Red Terror', the latter included atrocities that seemed particularly shocking to the strongly Catholic part of the Spanish population. Not only the murder of clergymen, monks and especially nuns, but also the exhumations of cadavers of long-dead religious people must be included here among the atrocities as prime example of humiliating treatment which greatly contributed to incense Conservative opinion against the Republicans. These reactions fuelled Franco's casting of his own insurgency against the Republicans as a 'Catholic Crusade'. Similar patterns, albeit on a smaller scale, can be found in the Greek Civil War (with a prelude in 1942–45, and the main conflict in 1946–49), where, as in Spain, the mutually inflicted atrocities left long-term scars on the collective national psyche.

Further examples of carefully planned atrocities can be found in Yugoslavia during the Second World War, where, infected by nationalism, Catholic Croats and Orthodox Serbs inflicted terrible casualties upon each other and on other ethnic minorities, and then again in the Yugoslav Wars in the 1990s, which were effectively wars of secession in which the central Serb-controlled government treated the secessionist states as rebels. Atrocities that were committed especially in the multi-religious Bosnia-Herzegovina generally had the backing of the political leaders of each group.[80] In both conflicts, we see especially the Croat but also the Serb parties infected with the racist ideas that had been spread to the Balkans by the propaganda machinery of the Third Reich if they had not already reached Yugoslavia as a part of the Central European *Zeitgeist* of the 1930s. It is not surprising in the light of the preponderant German ideology that there was little, if any, attempt by the Wehrmacht during the Second World War to win the hearts and minds of the populations of occupied territories, even though in the Ukraine, which was barely recovering from the Soviet-engineered Holodomor, the German armies were greeted initially as liberators from Soviet rule. The Wehrmacht fought the (increasingly Communist-led) insurgencies, which soon plagued them wherever they went, with punitive massacres.[81] In Yugoslavia, they were aided in their efforts by the Croat Fascist Ustaša, and in Poland by the anti-Communist Polish Home

Army.[82] The patterns of atrocities found here are fully in keeping with the racist ideologies held by the Germans and their allies.

Since the Second World War, we find evidence of planning in the atrocities committed in several countries of sub-Saharan Africa against populations who were portrayed as rebels against central authority, often due to tribal disputes over land, but also in the context of contested elections. Biafra has been mentioned; Kieran Mitton presents evidence in his article that atrocities committed by the RUF in Sierra Leone between 1991 and 2002 were premeditated. In Latin America, successive right-wing, often military, regimes reacted with great brutality against rebellious populations; notorious examples include the persecutions of Argentina in the 1970s and Salvador in the 1980s.[83] Russian repression of the insurgency and Islamic-*cum*-nationalist secession movement in Chechnya has produced similar patterns.[84] We can also find examples of premeditated atrocities committed against insurgents in the Middle East. Saddam Hussein's gassing of the Kurdish rebel population in his country's north eventually furnished one of the reasons that turned international public opinion against him. Reporting on the 2008 Gaza Conflict between Israelis and Palestinians, the UN-commissioned 'Goldstone Report' found that both sides had planned attacks on civilians.[85]

Spontaneous Atrocities

Apart from planned atrocities, throughout history, there have been clearly many examples of spontaneous atrocities. In a colonial warfare context, the massacre of the Sioux at Wounded Knee in December 1890 occurred spontaneously, escalating from a misunderstanding, just as that at Amritsar in 1919, which seems to have arisen from mistaking a religious gathering for a rebellion. The atrocities committed by American soldiers at My Lai in Vietnam in 1968 were only 'planned' insofar as the soldiers who flew into the village assumed that they would encounter enemy combatants, and that the civilian population would have left for the local market, and this was in part a horrible intelligence failure. The GIs were all geared up for battle with enemy combatants, but ended up killing defenceless people of all ages and both sexes.[86] Similarly, Bruno Reis and Pedro Oliveira show us that the Wiriyamu massacre was quite unpremeditated, again the outcome of intelligence failure and generally poor leadership. An exacerbating factor is clear that, whether endorsed or even planned by governments, or forbidden under pain of death, atrocities do occur. As with the crime of rape in civilian society, if the perpetrators are aware of criminality of the offence they have committed and the punishment awaiting them, they may be inclined to destroy the evidence, by killing the victims of rape or plunder or any non-fatal assault, thereby exacerbating their actions.

WHY DID SO MANY ATROCITIES OCCUR ON BOTH SIDES?

Theory and practice thus clearly diverged, even though some theoreticians – especially those dealing with 'recalcitrant' rebels or 'heretics' – give us a hint as to how ideological struggles could lead both sides to discard restraints and apply their full ideologically fuelled fury. The subject uniting the studies in this special issue is thus why and how atrocities occurred, even under liberal democracies, or regimes with a cultural background imbued with humanitarian ideals.

Let us turn, then, to existing explanations. One has already been alluded to and partly refuted at the beginning, namely that atrocities are due to the lack of professionalism and military discipline on the part of insurgents. First, examples abound in which the insurgents are led, or even largely consist of professional soldiers who had previously served in regular armies, as was the case for a fair proportion of the Spaniards in their *Guerrilla* against Napoleon's forces, or for the Irish rebels against Britain 1916–22. Second, the same lack of professionalism and military discipline can be diagnosed also among regular armed forces, and has indeed been one of the causes of atrocities there, as David Benest's article shows.

To turn to another set of answers, then, on the theoretical level, based on empirical findings, the path-breaking project of Hans-Henning Kortüm and his colleagues has posited a difference in the conduct of war in three different types of conflict (Type 1, and Types 2a and 2b), always *subjectively* perceived. The Kortüm typology divides wars, first, into *intra*cultural wars (wars that take place within one culture), in which contestants see each other as equally legitimate, in analogy to contestants in a two-sided game of sports. Second, it posits the category of trans-cultural wars, wars perceived as taking place between entities perceiving each other as quite different. Kortüm and his colleagues produce a pattern that shows the first category (*intra*cultural wars) to be fought to a much greater extent under restraints practised by both sides, and rules carefully enforced, while the second category, that of trans-cultural wars, is marked by a lack of restraints, and abounds with massacres and other atrocities. This second category, however, in this typology comprises two sub-categories: (1) *inter*-cultural wars, where, objectively, the cultures of the two sides in conflict are quite different (e.g. the Greek and Persian wars, the Barbarian invasions of the Roman Empire, the Arab and Turkish conquests of the East Roman Empire, or the Crusades); (2) subcultural wars, wars against an ostracised subculture within one's own population, especially heretics (and hence all the religious wars); and – of special interest to us here – opponents in civil wars, especially those where one side can be portrayed as rebels or insurgents against the legitimate state power. The hatred which arises from by ideological (including religious) differences within one's own culture against those who for ideological reasons form a subculture and challenge the legitimacy of the other parts of this culture, thus the hypothesis, tends to whip up emotions to the point of overriding natural and legal restraints on the prosecution of war.[87] All the patterns of pseudo-speciation which have been identified by Kai Erikson and others kick in: taboos which are generally upheld within one's own society in peace time, but even in war against opponents seen as equals, are cast aside on the basis of mutual and auto-suggestion that the adversary belongs to a different species, are sub-human, etc., thereby justifying even the killing of those entirely unable to fight, such as small children, the infirm and wounded, or very old people.[88]

This theory concerning the tendency towards extreme cruelty to be found in 'subcultural warfare' still allows for three different explanations of why atrocities occur in this context. One would be that, due to poor intelligence, surveillance and information flow, or other unintended factors, atrocities occur in a strictly unintentional way. This was clearly a very important factor in the Wiriyamu massacre examined by Pedro Oliveira and Bruno Reis in this issue, and also forms a feature of the My Lai massacre committed by American forces just 4 years earlier. Second, it might be due to unacceptably unprofessional leadership, for which both Jacques Frémeaux and David Benest present examples in their articles, where permissive attitudes on the part of commanding officers contribute to spontaneous and even planned atrocities. Benest sees this, in part at least, as a function of a 'broken society'. But the societies from which NATO forces that have committed atrocities in recent years come are hardly worse now than those of the 19th century. Is it due to the abandonment of religion? Some of the worst atrocities have been committed on behalf of a religion, mainly Christianity or Islam.

Yet another explanation for the occurrence of atrocities in insurgencies and COIN, notwithstanding the writings of theoreticians and practitioners cautioning against atrocities, as we have seen, is that these were often planned, and deliberately and systematically inflicted. Kieran Mitton's article on Sierra Leone offers very important insights here, as the Shame/Shameless Paradigm goes a long way to explain the participation of 'ordinary men' – here, indeed, children – in planned atrocities, especially where they serve to tie new members to the band of those perpetrating them, and to sever their natural bonds with their families, tribes and civil society. Similar patterns were found in Yugoslavia, among adult men, whose loyalty to a group of perpetrators of atrocities was secured by 'initiation' killings and mutilations which they were forced to conduct. Once they had been bloodied in this way, they could hardly give themselves up (and turn on the others), as the crimes they themselves had committed put them beyond any pardon. This leaves the question as to why the leaders of these child soldiers in Sierra Leone act the way they do. More classical interpretations – the greed and hatred he cites – as well as the tendency of a small pathological group among all human brutality are likely to provide at least part of the answer.

Mitton's article incidentally raises some very disturbing questions about morality by furnishing evidence that child soldiers can be brutalised to a degree that they become incapable of seeing the wrong in killing, maiming and torturing defenceless civilians, including children even younger than themselves. This challenges the proscriptions on such random killings in all major religions, and challenges our notions of natural ethics inherited from the Judaeo-Christian tradition which attributes to all (sane) human beings the capacity to know in their hearts, at least roughly, what is right and what is wrong. It equally challenges some key findings of evolutionary biology, ethics, and 'neuroethics', a part of cognitive science, where recent research seems to indicate that some moral dilemmas – not to mention the rarely contested taboo on killing young children – are resolved in similar fashions by people of different ages, cultures and genders.[89]

Accidents, intelligence failure and mismanagement are our chief keys to understanding why liberal democracies have committed atrocities, as in the case of the My Lai massacre in the Vietnam War, or those committed by British forces listed by David Benest, and not only authoritarian regimes (as in the Wiriyamu massacre analysed by Bruno Reis and Pedro Oliveira). By contrast, it is only logical that we find in racist, authoritarian and totalitarian Communist cultures the ideological contempt for 'inferior races', and indeed for the life of individual humans creating permissive mind-frames for atrocities to occur against non-combatant populations who supported rebels or opposition groups.[90]

Another factor is that of a spiral of violence and atrocities in which the abandonment of restraint by either side pushes the other – or is used as an excuse by the other – in turn to abandon restraints in the practice of violence. A classic quotation is that of Bugeaud, who in 1840 addressed the Assemblée Nationale's concerns about atrocities that were being committed in Algeria in the name of France. He urged them 'always to put France's interests above any absurd philanthropy for foreigners who behead our wounded and captured soldiers'.[91]

A further explanation is furnished by Benjamin Valentino, Paul Huth and Dylan Balch-Lindsay. On the basis of their quantitative study of mass killings of 50,000 or more civilians by governments in the contexts of guerrilla insurgencies have found the following of their hypotheses quantitatively confirmed by the dozens of case studies they have examined: governments are *not* more likely to resort to mass killings among civilians suspected of supporting the militants in ethnic conflicts than they are in ideological/political conflicts. Unsurprisingly, by contrast, they have found that such government-ordered mass killings are more likely to occur, the stronger the guerrilla forces are getting, the more support they are getting from the civilian population, the more threatened the regime feels and the longer the conflict drags on.[92] While our case studies of atrocities committed by western liberal regimes fortunately do not contain examples of massacres of such large numbers of victims, the overall pattern of these findings fits: namely the tendency to abandon restraints and take extreme measures the more desperate one gets.

We conclude this brief survey of our findings with the question of how to judge and punish atrocities in insurgencies and counterinsurgencies. This is clearly an area where the legal definition of war has to be widened to include not just the classic case for which international law was constructed, namely interstate war between sovereign entities. Any party that has actually committed atrocities, or parties that have planned them, clearly deserves to be prosecuted for war crimes. As Dale Walton shows in his contribution, 'world opinion', including that in liberal democracies, is now more critical of liberal democracies if any atrocities are committed by their soldiers, whether by accident or because of poor leadership at a lower level. Yet, as he rightly argues, this does not in any way 'excuse' other or state- or non-state actors who commit similar crimes, and the same attention must be given to those. Indeed, Valentino, Huth and Balch-Lindsay have found some support for their hypotheses that 'highly autocratic regimes are more likely to engage in mass killing during armed conflicts than highly democratic regimes'.[93] It would help greatly if all states signed up to the Rome Statute of the International Criminal Court, so that the possibility of prosecution for war crimes could help penetrate military (and political) institutional cultures worldwide to the point where soldiers of all countries have it drilled into them by their officers that certain actions are unacceptable, to counter the permissive culture which has occasionally thrived even in Liberal democracies under poor leadership, as David Benest's article has shown.

This special issue is thus not designed as a Liberal manifestation of self-flagellation, nor does it follow Charles Callwell's view of the world as divided into a 'savage' and a 'civilised' part, but tries to be equal handed in its approach. At best, we hope to have made a modest but honest contribution to our understanding of mechanisms that can trigger events which at least *our* culture condemns, and which we must learn to forestall. At worst, the case studies we present in this article will serve to enrich the data base for a more comprehensive study of atrocities in insurgencies and counterinsurgencies.

NOTES

1. 'Rebellion' is the politically most loaded term, as it has connotations of a legitimate government as the force of order on the one hand, and illegitimate rebels who challenge this order, in a Hobbesian tradition assumed to be better than chaos.
2. 'Common Article 3 of the Geneva Conventions of 1949' in Adam Roberts and Richard Guelff (eds) *Documents on the Laws of War* (3rd ed.) (Oxford: Oxford University Press 2000) p.302.
3. Erik Melander, Magnus Öberg and Jonathan Hall, 'Are "New Wars" More Atrocious? Battle Severity, Civilians Killed and Forced Migration before and after the End of the Cold War', *European Journal of International Relations* 15/3 (2009) pp.505–36.
4. Sextus Iulius Frontinus (born c. A.D. 35, d. A.D. 103 or 104), *Stratagematon* (between A.D. 84–96), II.xi.5–7, and IV.iii. 14 and iv.1, trans. and ed. by Charles E. Bennett, *Frontinus: The Stratagems and the Aqueducts of Rome* (London: William Heinemann for Loeb 1925) pp.192f, 294f. Admittedly, other examples he listed made the opposite argument, namely that enemies might be subdued through starvation or brutality.
5. Onosander, *Strategikos* (before A.D. 57) trans. by William A. Oldfather, *Aeneas Tacticus, Asclepiodotus, Onosander* (London: William Heinemann for Loeb 1923), Book XLII, pp.500–27.
6. Bertrand de Loque, pseudo. for François de Saillans, *Deux Traités: l'un de la guerre, l'autre du duel* (Lyon: Iacob Rayoyre 1589) in Beatrice Heuser (trans. and ed.), *The Strategy Makers: Thoughts on War and Society from Machiavelli to Clausewitz* (Santa Barbara, CA: Praeger 2010) p.55f.
7. Raymond de Beccarie de Pavie, Baron de Fourquevaux, *Instructions sur le faict de la Guerre* (Paris: M. Vascosan 1548) p.265.
8. Niccolò Machiavelli, *I Discorsi* (Florence: Bernardo di Giunta 1531); trans. by L. J. Walker SJ and Brian Richardson, *The Discourses* (London: Penguin 1998) pp.353–56.
9. Ibid., p.349.
10. Count Giacomo/Jacopo di Porcia, *Clarissimi viri Jacobi Pvrliliarvm comitis de re militaris liber* (Venice: Joannis Tacuinus di Tridino 1530), chapter 138, and see also 128, 133, 146, 147, 182.
11. Matthew Sutcliffe, *The Practice, Proceedings and Lawes of Armes* (London: Deputies of C. Barker 1593), p.204f; excerpts in Heuser (note 6) pp.62–86, esp. p.80. Sutcliffe attributes this to Sallust, but the reference is unclear.
12. Sutcliffe (note 11) p.207.
13. See, for example, Henry Humphry Evans Lloyd, 'Continuation of the History of the Late War in Germany, between the King of Prussia, and the Empress of Germany and Her Allies, Part II (1781)' in Patrick Speelman (ed.) *War, Society and Enlightenment: The Works of General Lloyd* (Leiden/Boston, MA: Brill 2005), p.453.
14. John Nigel Figgis, *Political Thought from Gerson to Grotius, 1414–1625* (1916; new ed. Harper: Hamilton 1960); Jonathan Scott, 'The Law of War: Grotius, Sidney, Locke and the Political Theory of Rebellion', *History of Political Thought* 13/4 (1992) pp.565–85.
15. Porcia (note 10) chapters 45, 49; see also chapters 118, 155.
16. Bernardino Rocca Piacentino, *De discorsi di gverra Libri quattro* (Venice: Damiano Denaro 1582) pp.165, 179.

17. Robert Barret, *The Theorike and Practike of Moderne Warres* (London: William Ponsonby 1598) p.11f.
18. Sutcliffe (note 11) p.11.
19. Sancho de Lonoño, *El discvrso sobre la forma de redvcir la Disciplina militar, à meyor y antigvo estado* (Brussels: 1589 new edition Madrid: Ministerio de Defensa 1992), quoted in Sutcliffe (note 11) p.12.
20. Sutcliffe (note 11) p.318.
21. Ibid., p.338.
22. Alberico Gentili, *De Iure Belli Libri Tres*, ed. and trans. by J. C. Rolfe (Oxford: Clarendon Press 1933) vol. I, p.511.
23. D. Bernardino de Mendoza,, *Teórica y práctica de la guerra* (Madrid: Pedro Madrigal 1595); new impression (Madrid: Ministerio de Defensa 1998) p.29f., excerpts in Heuser (note 6).
24. Bartolomé de las Casas, *Brevísima relación de la destrucción de las Indias* (1552), trans. by Andrew Hurley: *An Account, Much Abbreviated, of the Destruction of the Indies* (Indianapolis, IN: Hackett 2003).
25. Marquess Anibale Porroni, *Tratado Universal Militar Moderno* (Venetia: Francesco Nicolini 1676) p.417.
26. Paul Hay du Chastelet, *Traité de la guerre, ou politique militaire* (Paris: J. Guignard 1667) in Heuser (note 6) pp.117–23.
27. Don Alvaro de Navia Ossorio y Vigil, Vizconde de Puerto, and Marques de Santa Cruz de Marcenado, 'Reflexiones Militares (1724–30)' in Heuser (note 6) pp.135–43.
28. J. F. A. Le Mière de Corvey, *Des Partisans et des corps irréguliers* (Paris: Anselin and Pochard 1823) p.vii.
29. Beatrice Heuser, *The Evolution of Strategy: Thinking War from Antiquity to the Present* (Cambridge University Press 2010) pp.113–200.
30. Thomas Robert Bugeaud de la Piconnerie, Duc d'Isly, *Par l'épée et par la Charrue: Ecrits et Discours* in General Paul Azan (ed.) (Paris: PUF 1948) pp.110–18.
31. 'Proclamation de Bugeaud... juillet 1845' in Jacques Frémeaux (ed.) *Les Bureaux arabes dans l'Algérie de la Conquête* (Paris: Denoël 1993) p.294. See also Thomas Rid, 'The Nineteenth-Century Origins of Counterinsurgency Doctrine', *Journal of Strategic Studies* 33/5 (2010) pp.727–58.
32. Marshal Louis Hubert Gonzalve Lyautey, *Rayonnement* (writings ed. by Patrick Heidsieck) (Montreal: Gallimard/Editions Variétés 1944).
33. Jacques Frémeaux, *Internvention et Humanisme: Le style des armées françaises en Afrique au XIXe siècle* (Paris: Economica 2006) p.67.
34. Charles E. Callwell, *Small Wars, Their Principles and Practice* (3rd ed.) (London: HMSO [1899] 1906) p.40.
35. Quoted in Brian Lapping, *End of Empire* (London: Granada 1985) p.224.
36. Sir Robert Thompson, *Defeating Communist Insurgency: Experiences from Malaya and Vietnam* (London: Chatto & Windus 1966).
37. Gen. David Petraeus and Gen. James Amos (eds), *The US Army/Marine Corps Counterinsurgency Manual* (Chicago, IL: University of Chicago Press 2007); see also Beatrice Heuser, 'The Cultural Revolution in Counterinsurgency', *Journal of Strategic Studies* 30/1 (2007) pp.153–71.
38. Carl von Clausewitz, 'Bekenntnisdenkschrift von 1812' in Werner Hahlweg (ed.) *Carl von Clausewitz: Schriften - Aufsätze - Studien - Briefe* (Göttingen: Vandenhoeck & Ruprecht 1966) vol.1, pp.730–34; Carl von Clausewitz, *On War* (trans. and ed. by Michael Howard and Peter Paret) (Princeton, NJ: Princeton University Press 1976) Book VI.
39. General Wojciech Chrzanowski, *Ueber den Partheigaenger-Krieg: Eine Skizze* trans. from Polish (Berlin: Stuhrsche Buchhandlung 1846).
40. Ernesto 'Che' Guevara, '¿Qué es un guerrillero?', *Revolución* (Feb. 1959), online at <http://www.marxists.org/espanol/guevara/59-quees.htm>, accessed 19 Jan. 2012.
41. Beatrice Heuser, 'Small Wars in the Age of Clausewitz: Watershed between Partisan War and People's War', *Journal of Strategic Studies* 33/1 (2010) pp.137–60.
42. Johann Ewald, *Treatise on Partisan Warfare* (1785), ed. and trans. by Robert A. Selig and David Curtis Skaggs (New York: Greenwood Press 1991) p.69.
43. Georg Wilhelm Freiherr von Valentini, *Abhandlung über den kleinen Krieg und über den Gebrauch der leichten Truppen mit Rücksicht auf den französischen Krieg* (4th ed.) (Berlin: J.W. Boicke [1799] 1820).

44. Heinrich von Brandt, *Ueber Spanien mit besonderer Hinsicht auf einen etwaigen Krieg* (Berlin: Schüppel'sche Buchhandlung 1823), trans. by a British officer *The Two Minas and the Spanish Guerrillas* (London: T. Egerton 1825) p.54.

45. José Maria de Carvajal, *Reglamento para las Pardidas de Guerrilla* (Cadiz: Oficina de Don Nicolas Gomez de Requena, 11 Jun. 1812) pp.17–20.

46. Don Francisco de Copons y Navia, *Diario de las Operaciones de la Division expedicionaria al mando del Mariscal de Campo Don Francisco de Copons y Navia, desde su salida de Cadiz en el mes de Octubre de 1811, hasta que regreso en Marzo de 1812, despues de haber desendido la plaza de Tarifa* (Vich [now Vic]: Antonio Brusi 1814) p.8.

47. Hubert Heyries, 'La conception garibaldienne de la guérila: Armée de la Nation ou Nation armée?' in Centre d'histoire et de prospective militaires (ed.) *Guerre civile – guerilla – terrorisme: entre hier et avenir* (Pully, Switzerland: 1998) vol. II, pp.231–44.

48. *Les Guérillas: Instructions pour les volontaires Francs-Tireurs et Mobiles de l'Armée des Vosges* (Dôle: de Pillot 1870) pp.2–30 passim.

49. Mao Tse-tung, 'Yu Chi Chan (1937)', trans. by Samuel B. Griffith, *On Guerrilla Warfare*. (New York: Praeger 1961, reprinted Champaign, IL: University of Illinois Press 2000) p.92.

50. Quoted in General Vo Nguyen Giap, *People's War, People's Army: The Viet Cong Insurrection Manual* (Dehra Dun, India: Natraj Publishers 1974) p.v–vi.

51. Vietnam. Nha kê-hoach tâm-lý-chiên, *Viet Cong Atrocities and Sabotage in South Vietnam* ([presumed to be Saigon]: Ministry of Information 1967).

52. Ernesto 'Che' Guevara, 'Guerrilla Warfare' (1960) in Che Guevara (ed.) *Guerrilla Warfare*, trans. by Anon. (London: Souvenir Press 2003) pp.13, 24f.

53. Régis Debray, *La Guérilla du Che* (Paris: Eds du Seuil 1974).

54. The best historical overviews are given in John Ellis, *A Short History of Guerrilla Warfare* (London: Ian Allan 1975); Werner Hahlweg, *Guerilla: Krieg ohne Fronten* (Stuttgart: Kohlhammer 1968), and for more recent times, Ian Frederick William Beckett, *Modern Insurgencies and Counter-Insurgencies: Guerrillas and Their Opponents since 1750* (London: Routledge 2001).

55. German Schleifheim von Sulsfort [Hans Jakob Christoffel von Grimmelshausen], *Der abenteuerliche Simplicissimus* (Mompelgart [Nuremberg]: Fillion [Müller] 1669) passim.

56. See Deuteronomy 20.13f; and 1 Samuel 15.2f.

57. See, for example, the Melians, who were treated thus after their unconditional surrender to Athens, whose demand to become allies they had defied, in Thucydides, *History of the Peloponnesian War* V.116 (before 400 BCE), trans. by Rex Warner (London: Penguin 1954) p.408.

58. Hans van Wees, *Greek Warfare, Myths and Realities* (London: Duckworth 2004) p.135f.

59. Caesar, *De Bello Gallico*, VII.28.4 and VII.47.5.

60. Michael Prestwich, 'Transcultural Warfare: The Later Middle Ages' in Hans-Henning Kortüm (ed.) *Transcultural Wars from the Middle Ages to the 21st Century* (Berlin: Academie-Verlag 2006) pp.43–56.

61. For example, Gregory of Tours, *The History of the Franks* VI.31, trans. by Lewis Thorpe (Harmondsworth: Penguin 1974) pp.359–61.

62. Bernhard Zeller, 'Collective Identities, War and Integration in the Early Middle Ages' in Anja V. Hartmann and Beatrice Heuser (eds) *War, Peace and World Orders in European History* (London: Routledge 2001) pp.102–12.

63. Matthew Strickland, 'Rules of War or War without Rules? Some Reflections on Conduct and the Treatment of Non-Combatants in Medieval Transcultural Wars' in Kortüm (note 60) p.123.

64. Ibid., p.126.

65. Arthur Jones (ed. and trans.), *The History of Gruffydd ap Cynan* (Manchester University Press 1910) S.141.

66. Quoted in Strickland (note 63) p.125.

67. *Gesta Stephani*, quoted in John Ellis (note 54) p.40.

68. Anne Curry, 'War, Peace and National Identity in the Hundred Years' War' in Hartmann and Heuser (note 62) pp.141–54.

69. R.J. Knecht: *The French Wars of Religion* (2nd ed.) (Harlow: Longman 1996) pp.39–53; see also Philip Benedict, 'The Saint Bartholomew's Massacres in the Provinces', *The Historical Journal* 21/2 (1978) pp.205–25.

70. Aline Goosens, 'Wars of Religion: The Examples of France, Spain and the Low Countries in the 16th Century' in Hartmann and Heuser (note 62) pp.160–74.

71. Johann Philip Abeln, *Theatrum Europaeum* (Frankfurt am Main: Matthaeus Merian 1646) vol. 2, Entry for 1631, p.368.
72. Barbara Donagan, 'Atrocity, War Crime, and Treason in the English Civil War', *The American Historical Review* 99/4 (1994) pp.1137–66; Frank Tallett, 'Barbarism in War: Soldiers and Civilians in the British Isles, c. 1641–1652' in G. Kassimoris (ed.) *Warriors Dishonour: Barbarity, Morality and Torture in Modern Warfare* (Aldershot: Ashgate 2006) pp.19–39.
73. Gwynn Lewis, 'The White Terror of 1815 in the Département du Gard', *Past & Present* 58 (1973) pp.108–35.
74. Daniel Krebs, 'The Making of Prisoners of War: Rituals of Surrender in the American War of Independence, 1776–1783', *Militärgeschichtliche Zeitschrift* 2 (2004) pp.1–25.
75. John Prebble, *Culloden* (London: Secker and Warburg 1961), passim; Hugh Gough, 'Genocide and the Bicentenary: The French Revolution and the Revenge of the Vendée', *The Historical Journal* 30/4 (1987) pp.977–88; Michael Wagner, '"Normalkrieg" oder "Völkermord"? Neue Forschungen zur Niederwerfung des Aufstandes in der Vendée', *Francia* 22/2 (1995) pp.177–85; see also Hew Strachan, 'A General Typology of Transcultural Wars: The Modern Age' in Kortüm (note 60) p.89f.
76. See Jacques Frémeaux, *De quoi fut fait l'empire: Les guerres coloniales au XIXe siècle* (Paris: CNRS Editions 2009).
77. Beatrice Heuser, 'Misleading Paradigms of War: States and Non-State Actors, Combatants and Non-Combatants', *War and Society* 27/2 (2008) pp.1–24.
78. Martin Ewans, *European Atrocity, African Catastrophe: Leopold II, the Congo Free State and Its Aftermath* (London: Routledge Curzon 2002) pp.164, 177.
79. Horst Drechsler, *Aufstände in Südwestafrika: Der Kampf der Herero und Nama 1904 bis 1907 gegen die deutsche Kolonialherrschaft* (East Berlin: Dietz 1984); Jürgen Zimmerer and Joachim Zeller (eds) *Genocide in German South-West Africa*, trans. from German by Edward Neather (Monmouth: Merlin 2008).
80. See, for example, Jan Willem Honig and Norbert Both, *Srebrenica: Record of a War Crime* (London: Penguin 1996).
81. Peter Lieb, *Konventioneller Krieg oder NS-Weltanschauungskrieg? Kriegführung und Partisanenbekämpfung in Frankreich 1943/44* (München: Oldenbourg 2007).
82. Jonathan Gumz, 'Wehrmacht Perceptions of Mass Violence in Croatia, 1941–1942', *The Historical Journal* 44/4 (2001) pp.1015–38; Klaus Schmider, *Partisanenkrieg in Jugoslawien 1941–1944* (Hamburg: E.S. Mittler & Sohn 2002); Bernhard Chiari (ed.) *Die polnische Heimatarmee: Geschichte und Mythos der Armija Krajowa* (Munich: Oldenbourg 2003).
83. Nicholas Robins, *Native Insurgencies and the Genocidal Impulse in the Americas* (Bloomington, IN: Indiana University Press 2005).
84. See, e.g. 'Russia/Chechnya', Human Rights Watch 12/5 (D) (April 2000).
85. 'Human Rights in Palestine and Other Occupied Arab Territories: Report of the United Nations Fact Finding Mission on the Gaza Conflict', online at<www2.ohchr.org/english/bodies/hrcouncil/specialsession/9/docs/UNFFMGC_Report.pdf.>, accessed 5 Apr. 2011.
86. Michael Bilton and Kevin Sim, *Four Hours in My Lai: A War Crime and Its Aftermath* (London: Viking 1992); Bernd Greiner, *War without Fronts: The Americans in Vietnam* trans. by Anne Wyburd and Victoria Fern (London: Bodley Head 2009).
87. Kortüm (note 60) pp.11–26.
88. See Kai Erikson, 'On Pseudospeciation and Social Speciation' in Charles B. Strozier and Michel Flynn (eds) *Genocide, War and Human Survival* (Lanham, MD: Rowman and Littlefield 1996) pp.51–57.
89. See the work of Marc Hauser, *Moral Minds* (New York: Ecco 2006), and John Mikhail, 'Universal Moral Grammar: Theory, Evidence and the Future', *Trends in Cognitive Sciences* 11/4 (2007) pp.143–52.
90. Mathew Cooper, *The Phantom War: The German Struggle against Soviet Partisans 1941–1944* (London: Macdonald 1979); Philip W. Blood, *Hitler's Bandit Hunters: The SS and the Nazi Occupation of Europe* (Washington, DC: Potomac Books 2006); Christian Hartmann, Johannes Hürter, Peter Lieb and Dieter Pohl, *Der deutsche Krieg im Osten 1941–1944: Facetten einer Grenzüberschreitung* (Munich: Oldenbourg 2009).
91. Bugeaud de la Piconnerie (note 30) p.67f.
92. Benjamin Valentino, Paul Huth and Dylan Balch-Lindsay, '"Draining the Sea": Mass Killing and Guerrilla Warfare', *International Organization* 58/2 (2004) pp.375–407.
93. Ibid., pp.400–7.

SUGGESTED FURTHER READING FROM *CIVIL WARS*, 14(1), MARCH 2012:

- British Atrocities in Counterinsurgency – *David Benest*
- The French Experience in Algeria: Doctrine, Violence, and Lessons Learnt – *Jacques Frémeaux*
- From Retaliation to Open Bridges: Moshe Dayan's Evolving Approach towards the Population in Counter Insurgency – *Eitan Shamir*
- Cutting heads or winning hearts Portuguese late colonial counterinsurgency and the Wiriyamu massacre of 1972 – *Bruno C. Reis & Pedro A. Oliveira*
- Irrational Actors and The Process of Brutalisation: Understanding Atrocity In The Sierra Leonean Conflict (1991–2002) – *Kieran Mitton*
- Victory through Villianization: Atrocity, Global Opinion, and Insurgent Strategic Advantage – *C. Dale Walton*

Lessons learnt? Cultural transfer and revolutionary wars, 1775–1831

Beatrice Heuser

Department of Politics & IR, University of Reading, Reading, UK

Did participants in small wars in the period 1775–1831 learn from previous or contemporary examples? While this is difficult to prove for participants who left no written records, there is considerable evidence in existing publications by practitioners that they did indeed draw out lessons from recent insurgencies, either from their own experience or from events elsewhere which they studied from afar, especially the Spanish Guerrilla, which had already become legendary. Most authors showed an interest in how to stage insurgencies rather than in how to quell them. Even then, transfer did not come in a package of tactics-cum-values, but in each case in different configurations.

Revolutions in Warfare, 1775–1831

The period 1775–1831 saw not only political revolutions on both sides of the Atlantic, but also revolutions in warfare. Revolutions, in the plural: first, it ushered in what Clausewitz called 'war in its absolute perfection', Napoleon's wars,[1] the French waging of which was sometimes described as 'small war in big style'.[2] Secondly, French warfare under Napoleon could also be seen as having lost the special operations element of 'small war' discussed by Bertrand Fonck and George Satterfield: the Hanoverian General Georg von Berenhorst noted that in the Napoleonic Wars, 'major war has swallowed small war, and in the campaigns of 1805–1806 there was no small war.'[3] Thirdly, the period 1775–1831 also saw the partial replacement of the '*petite guerre*', what we would call today special operations performed by special professional forces acting alongside regular armies, into '*guerrilla*', ostensibly the same word, only this time in Spanish, but encapsulating for us today the element of insurgency, a politically motivated uprising.[4]

The series of uprisings considered here began with the American War of Independence (1775–1783). This is of course not to deny the origins of some features of this war in the French and Indian Wars, which drew heavily on tactics and traditions of the Native Americans. But those earlier wars lacked the element of insurgency.

The American War of Independence was followed by uprisings essentially opposed to the changes introduced by the French Revolution, or simply against French occupying forces: the insurgencies in the Vendée and Brittany (the *Chouannerie*, 1793–1800), the Spanish War of Independence (1808–1814), a series of German anti-French uprisings (1808–1813), the Tyrolean uprising against the Bavarians (who had secured this area as France's allies in the wake of joint victories; 1809); the irregular warfare of Russians, acting alongside the Imperial Russian Army (1812–1813); and the German Wars of Liberation (1813–1815). These two groups – the American War of Independence on the one hand and the anti-revolutionary/anti-French wars on the other – seem for us today to belong to two different categories: progressive and reactionary uprisings. For contemporaries, however, all these uprisings would have resembled each other in that they were bids for freedom and self-determination. This would unite them with a later wave of insurgencies or wars of independence – the Greek War of Independence (1821–1832), the unsuccessful Polish uprising of 1830–1831, and the long Italian struggle for independence and national unity.

For all those who, following Guibert, Lloyd, Kant, and Clausewitz, see a link between the structure and values of a society and the way it wages war, the question arises, first, as to the extent to which this rebel spirit was shared in all these insurgencies affecting warfare. And secondly, was there a learning process, a transfer of tactics, and perhaps also of ideals of liberty and self-determination, from any of these wars to later ones?

The American War of Independence and its influence on European authors

It is no news that European liberals – Britons apart – looked with sympathy upon the American attempts to realise these shared values in the creation of their new state. The French *lumières* whose country had lost out in the struggle for possession of the New World found it easy to extend their approval. As F.E. Toulongeon, earliest biographer of Guibert, noted even in 1803, 'America was the great show of a people seizing its usurped rights; she fought for her liberty; and ... France ... had decided to help her, which would bring [America] great lessons and great examples.'[5]

The flow of ideas and even tactics from Europe to America is thus not in doubt. But was there also a flow in the opposite direction, from America to Europe? Peter Paret found little evidence of this in the writings of any participants.[6] By contrast, Robert A. Selig and David Curtis Skaggs have found indications of such a transfer, at least in terms of values.[7] This question merits a closer look.

The American War of Independence saw classic set-piece battles between professional soldiers fighting for the British crown and American professional soldiers. It also saw local militias harassing the British forces through ambushes and skirmishes making optimal use of their own knowledge of the terrain. The militia would usually prove inferior to the professional forces if confronted with them in open battle. As long as they avoided these, however, they could inflict substantial casualties on the other side. Time and again, they attacked the British forces along their routes, particularly in forested or marshy areas. Several American partisans,[8] such as Daniel Morgan and Andrew Pickens, had learnt their trade in the 'French and Indian Wars', and were largely inspired by the practices of Native Americans.[9] General Nathaniel Greene formed irregular units, one of which was led by 'Swampfox' Francis Marion, whose mission was to harass the redcoats in the swamps of West Carolina. The American parties would attack isolated British positions or lure them into dangerous areas where the vanguard and rearguard could be easily isolated and attacked. Greene thus described his tactics:

> I have been obliged to practice that by finesse which I dared not attempt by force.... There are few generals that has [sic] run oftener or more lustily than I have done... But I have taken care not to run too far, and commonly I have run as fast forward as backward, to convince our Enemy that we were as fast forward as backward,... that we were like a Crab, that could run either way.... We fight, get beat, rise and fight again.[10]

These partisans used tactics which were essentially not different from those used by the purely mercenary professional partisans that we have encountered in the article by George Satterfield and Bertrand Fonck. They were, however, inspired by a political cause, which was that of their own independence from London's tutelage.

Morgan, Pickens, Greene, and Marion faced not only the British regular forces, but also other partisans: King George III of England, Scotland, and Ireland, Elector of Hanover, had famously imported into the theatre of war soldiers 'bought' from other German princes, especially the famous Hessians of Landgrave Frederick II of Hesse-Kassel. These men were largely pressed into service, but some, serving in 'Hunter' (*Jäger*) units, were specialists in small war who had already fought in the Seven Years War. Among them was the former forester or forest warden (*Förster*) Andreas Emmerich (1739–1809), whose professional background is significant for the entire way of war of these special units that would later become known as 'rangers' in English.

Emmerich had served in a *Jäger*-unit of the Duke of Brunswick, before being appointed master forester by Frederick II. This seems not to have fulfilled him, and he signed up alongside the Hanoverians who went to fight for their Elector in America. In 1789, drawing on his experiences in the Seven Years War and the American War of Independence, he published a work on partisan warfare which did not differ much from the earlier standard works on the fighting of these irregular units by Grandmaison and Jeney.[11] Perhaps it seemed to him – as for Machiavelli and many others before him – that the genre of a military manual was not the place to discuss politics. His subsequent actions, however, are evidence of a change of heart that he had undergone in America, as he does not seem to have enlisted originally out of any ideological motivations. In 1809, 70 years old, impoverished and frustrated in his attempts to secure a pension from one of his previous employers, he turned rebel himself. Together with Colonel William von Dörnberg, he organised an insurgency against Jérôme, Napoleon's brother who had been installed by the Emperor as King of Westphalia (which included much of the former Landgrave of Hesse-Kassel's territories). The insurgents were defeated, but while Dörnberg managed to escape, Emmerich was arrested, tried, and executed in Kassel.[12] Parallels between the motivations of this insurgency and those of the American rebels are obvious: they all aimed to shake off what they saw as yoke of foreign oppression.

Another Hessian who had fought in America in the service of the Hanoverians, again as commander of a *Jäger* unit, was Johann Ewald (1744–1813). Back from the American war, Ewald joined the Danish army, received a series of promotions, and was finally ennobled.[13] On the basis of his experiences in Europe and America, he wrote a whole series of treatises on small war and light troops.[14] At best one can identify here a certain respect for the adversary, his morale, his tactics. Ewald stressed the importance of maintaining good relations with the indigenous population as they would be terrible to have as enemies, if made desperate.[15] The preface of his *Treatise on Small War*, dealing with the relation between officers and men within the armed forces, also testifies to an enlightened attitude to the very soldiers he commanded:

> The soldier is expected to do his duty and to know what he has been taught. Can one not address the same requirement at his [commanding] officer, who in all fortunate outcomes of the war earns all the laurels, while the soldier has to pay for the officer's ignorance with his life, his health or his freedom? . . . Do you not believe that the blood of those, who are sacrificed due to ignorance, will scream for the revenge to the Great Judge? . . . Is it not our duty [as officers] to learn, when peace gives us more than enough leisure to do so, so that we will be free from such accusations in times of war?[16]

Was this evidence of a value transfer? Ewald may have been influenced by the egalitarian values of the American Revolution, or at any rate despite having fought on the British side in that conflict, he was inspired by ideas of the Enlightenment, of which the American Revolution itself was born. What we can say is that his writing contrasts with those of other Hessian and Prussian authors who had not seen the New World, and which show no signs of a new spirit.[17] Was this coincidence?

In Prussia, soon thereafter, the writing of August Count Neidhardt von Gneisenau (1760–1831) stands out among those of his peers. He had been sent to the American war by the Markgrave Charles-Alexander of Brandenburg-Ansbach to fight alongside the Hanoverians. One might suppose that this experience influenced him, like Emmerich and Ewald, perhaps not to go quite so far as to become a republican, but to respect the individual soldiers more and to concede that they had a personal dignity, and to call for freedom from all foreign occupation. He laid down plans for a general Prussian uprising against Napoleon, in analogy to America's uprising against British rule (but he explicitly evoked the Spanish uprising against Napoleon).[18]

Another Central European, the Pole Tadeusz Kościusko (1746–1817), was himself a vehicle of cultural transfer in both directions. Initially educated as an engineer in the Polish army, he completed his military education in France in 1769. After his country had been partitioned by Prussia, Austria, and Russia in 1772, and while Kościusko was still in France, he was invited by the American rebels to join their side. Following their call, he became an aide-de-camp to General Washington, and was then involved in the construction of the West Point Military Academy, and is counted among the heroes of the American War of Independence. Having returned to his own country, he tried unsuccessfully in 1791–1794 to help organise the resistance of Poland against the second partition. He was taken prisoner by the Russians. Upon his liberation he went back, first to America, then to France where in the village where he had settled near Fontainebleau, he staved off a rampage of pillaging by the anti-Napoleonic coalition forces in 1814. As a septuagenarian, he managed to undertake a final trip to his native Poland where he freed the serfs on his family domains. He died in Switzerland.[19] Jointly with Józef Pawlikowski, Kościusko authored a work which was published anonymously in 1800 under the title *Can the Poles achieve their independence?*[20] This not only explored possibilities of rising up against Russian occupation, but also pondered questions of national and political identity.[21]

There is thus some evidence of *reciprocal* – and not merely unilateral – influence between the American and the European revolutions (and of course we can only seize upon those who bequeathed their thoughts to us in writing).

The Vendée, Spain, and Napoleon's Russian campaign

We have clearer evidence of influences arising from the first *anti*-revolutionary uprisings, the uprising in the Vendée (1793–1796) and the Breton *Chouannerie*. The uprisings had started over the government's orders to the priests to take an oath on the new constitution. As Alan Forrest has shown in his contribution to this volume, the insurgents mixed aspects of *petite guerre* with regular battles against the French Revolutionary forces. The repression of these uprisings took a relatively humane form under Lazare Hoche, and a brutal form with his predecessors and successors. Those included General Turreau with his 'infernal columns', Jean-Antoine Rossignol, François-Sévérin Marceau, Jean-Baptiste Kléber, the Alsatian François-Joseph Westermann, and Jean-Pierre Travot. Westermann told the Revolutionary government, the Comité du Salut Public, on 23 December 1793: 'There no longer is a Vendée. It is dead under our free-ranging sabre, with its women and its children.... I have crushed the children under the hooves of the horses, massacred the women... I have exterminated everything.'[22] By contrast, in 1795 Hoche took the route of psychological persuasion, urging his troops not to 'devastate the humble cottage of the country-dweller over whom the two parties seem to tear each other to pieces', and to see 'the bayonet... only... as secondary means': 'we have to protect and shelter the weak and let the property of all be respected.' [23] In a proclamation of November 1795, he tried to appease the Vendeans with religious tolerance, claiming that 'the real Republicans do not commit atrocities' but 'want to give you the kiss of peace. They come to free you from tyranny and not to strangle you.'[24] Hoche ordered his soldiers to leave religious questions alone and to treat priests with respect.[25]

While Westermann was eventually guillotined for his excesses, Hoche was criticised by the Comité du Salut Public for his soft approach. Hoche's pacification attempts had no lasting success as the insurgency resumed in 1795. Nevertheless, his approach would become famous, being cited with approval by Charles Edward Callwell, the first British author on counter-insurgency, at the end of the nineteenth century. In Calwell's view, however, Hoche's humane approach would work only in 'civilised' Europe, not in the rest of the world.[26] Two East Germans, August Rühle von Lilienstern and Carl von Clausewitz, both in their teens when all this happened, would later comment only on the valour of the Vendeans;[27] more detailed information about these events was difficult to come by in Berlin at the time, as few foreigners witnessed them. The 'Choannerie' in Brittany was rarely, if ever, mentioned.

By contrast, the anti-Bavarian uprising in the Tyrol, a region taken from Austria by the victorious French and given to their Bavarian allies, was much reported throughout the German-speaking lands. As with the Prussian uprisings described by Martin Rink in his article in this issue, the Tyrolean uprisings did not gain much support from the Austrian leadership. Prince Metternich, the foreign minister, initially supported the Tyroleans, but soon saw the insurrection slip from his control, from which point he, too, backed those out to quash it. Equally, the Austrian Archduke Charles, one of the main commanders in the wars with Revolutionary France, feared that such a 'people's fight' ('*Volkskampf*') might be acceptable as a last resort, but generally had the disadvantage of undermining 'the prosperity and the morals of the people for a long period of time'.[28] Given the lack of backing from Vienna, the uprising was tragically unsuccessful, and would furnish material for poems and later for romantic films, but not models for insurgency literature.

Unlike the Tyrolean uprisings, the Spanish uprising against the French occupiers was an inspiration to all and sundry. It was both romanticised and ultimately successful, and it was reported widely as it was witnessed by soldiers from countries other than France and Spain. The French forces themselves were conscious of its particularities as they had other campaigns to compare this with. A young captain of the French hussars was inclined

> even to compare two types of absolutely different wars; the war of regular troops who normally take little interest in the quarrel which they support [through their fighting], and the war of resistance which a nation can put up against conquering regular armies [*armées de ligne*].... On the Spanish peninsula, we were hardly called upon to fight against regular troops which are everywhere much the same, but against a people.[29]

Generally it can be said that the classical narrative of the Spanish *Guerrilla*, as a purely ideologically driven uprising against a tyrannical foreign occupation regime, was already well established outside Spain in the 1820s and 1830s, as we shall see presently.[30] The French colluded with the Spanish in establishing this narrative, and it was eagerly taken up by other authors. For German and Polish authors in particular, it would be the great point of reference, for better or worse, of what they considered organising (and partly implemented) in their respective uprisings against France and Russia.[31]

Jomini and Clausewitz

The Swiss-born Henri Baron de Jomini (1779–1869), a general in Napoleon's armies and the most influential military writer of his times, had himself witnessed the Peninsular War. He ranked the Guerrilla as example of 'national' war among 'the most terrifying of all'. He cited the sudden disappearance of an entire French company, without any trace, and without ever being able to find out what had happened to them.[32] Among those conservatives who were particularly worried about the potential consequences of such uprisings, Jomini was perhaps the most sober. His critical attitude was fact-based and focused on the atrocities produced during the Spanish War of Independence, rather than on the nightmares of popular unrest which haunted all noblemen in Europe since the French Revolutionaries had guillotined their peers in large numbers. Jomini's writings would dominate the French, Russian, American, and to some extent even the British military world (and especially the educational establishments) of the nineteenth century, and thus underpinned scepticism towards any *guerrilla* activities.

Gneisenau's protégé, Carl von Clausewitz (1790–1831), by contrast, had not witnessed the Spanish War of Independence. Nevertheless, he was still strongly influenced by it. Taking up the terminology of the French military philosopher Guibert,[33] Clausewitz underscored that in Spain, war 'spontaneously became the concern of the people.'[34] The published works of Clausewitz, even the posthumous *On War*, were curiously devoid of reflections on values or political systems – curious, given the articulation of the link between war and politics for which he is most famous. Only in his private letters, especially one he wrote to Gneisenau in 1812, published long after his death, do we find political arguments for a military restructuring, which would never be put into practice.[35] But we search in vain for any reflections on the attitudes and values of the Americans in their struggle for independence from London, even if General Washington is mentioned in Clausewitz's writings.[36]

Clausewitz's views on warfare were also strongly influenced by his own experience of Napoleon's campaign of 1812, which he witnessed on the side of the Russians.[37] From this he deduced that in general defence was stronger than offence, and, based on this and against his second-hand information on the events in the Vendée, the Tyrol, and the Peninsular War, he formulated his famous concepts of 'people's war' (*Volkskrieg*) and the arming of the people

(*Volksbewaffnung*), later translated as 'the nation in arms', in Book VI of *On War*. Clausewitz wrongly claimed that only the nineteenth century had produced this phenomenon, supposedly unknown in civilised Europe until that date, of the organised arming of the population. [38] In reality, there had of course been many previous examples of religiously or politically inspired uprisings that had mobilised a large proportion of a population. Such were the sixteenth-century religious wars in France or the Eighty Years War of the Low Countries against the Spanish – wars Clausewitz simply ignored. The American War of Independence and the uprising of the Vendée had of course also furnished pre-nineteenth century examples of popular uprisings – ones that Clausewitz knew about as he demonstrated elsewhere in his writings.

Clausewitz himself later became an indirect victim of an insurgency. He died of cholera in a Prussian-occupied part of Poland during a preventative deployment to guard against a spread of the Polish insurgency of 1830–1831 in the Russian-occupied areas.

Bugeaud

Arguably the most important French witness of the Spanish Guerrilla was probably Thomas Robert Bugeaud, Marquis de la Piconnerie and Duke of Isly (1784–1849), as he transposed his Spanish experience and the French countermeasures to the Algerian theatre where he served in the 1830s, soon after the French conquest, and from 1840 to 1847 was governor general. In his personal letters, he showed no interest in either strategy or politics, and certainly no compassion for the Spanish,[39] as we can see from an account of the events of the *dos de Mayo* 1808, which he sent to his sister:

> The populace of Madrid took the fancy of rising up in revolt on 2 May. They threw themselves on isolated Frenchmen whom they strangled; then they moved to the arsenal, seized it, took out the cannon, equipped themselves with shotguns, and started to wage a small war in the streets with some French guards. On our side, we were not inactive. We drummed out general [alert], we ran about the town, and their success did not last long. We attacked them vigorously on the bridges. We toppled them over, seized their [artillery] pieces, and within an hour, the chaotic crowd disappeared. On the same day, we shot a good number of culprits. We have already lost some men in that operation. I came away with a bruise and a light strangulation. The insurgents wanted to strangle our soldiers in the hospital, but those [among the invalids] who were the strongest broke into the arms magazines and exterminated their assailants. Calm seems to have been re-established, but one must not rely on it... I... am not at ease walking through the streets: I always have my hand on my sword, as even before the revolt, there were daily assassinations, and those gents took our moderation for weakness.[40]

In 1808–1809 Bugeaud also took part in the second siege of 'that infernal Zaragoza', which became a symbol of French rather than Spanish cruelty. In February 1809, Bugeaud sent his sister the following account:

Even though we assailed their ramparts for a fortnight and held a part of the town, the inhabitants, incited by the hatred they feel for us, by the priests and the fanaticism, seem to want to bury themselves in their city according to the example of ancient Numantia.[41] They defend themselves with an incredible determination and make us pay dearly for our little victory... Each monastery, each house presents us with the same resistance as a fortress, and each has to be besieged separately. Each has to be fought over on foot, from the cellars to the loft, and it is only once one has killed everything [!] with thrusts of the bayonet or thrown everything [!] out of the windows that one can claim to be master of the house. One has hardly triumphed [in one house] when [somebody in] the neighbouring house throws grenades, missiles, and a hailstorm of gunshots at us through holes made for this purpose. You have to construct barricades, cover up quickly until measures have been taken to attack the new fortress, and one only manages by breaking down walls, as it is impossible to pass through the streets: the entire army would perish there in two hours.[42]

When Zaragoza surrendered on 20 February 1809, French losses amounted to 4000 men, while the city mourned the death of about 50,000 inhabitants. Bugeaud was promoted to captain and went on to pursue a brilliant military career. He applied equally radical repressive measures in Algeria, which he explained as follows:

It means little to cross the mountains and to fight the mountain-people once or twice; in order to subdue them, one must attack their interests. One can succeed by passing through like a dart; one has to come down heavily on the territory of each tribe; one must... stay long enough to destroy the villages, cut down the fruit trees, burn or uproot the harvest, empty the silos, search the gullies, rocks and caves so as to seize the women, the children, and the old; it is only thus that one can make the proud mountain people capitulate. If one limits oneself to checking only one or several roads, one only comes across the fighters, one fights them with more or less advantage, but one reaches neither the population, nor the riches, and the results will always be negative.[43]

Bugeaud thus showed not the slightest sympathy for insurgents, which is all the more surprising as he was, on his mother's side, a descendant of Irish Jacobite rebels who had been forced to flee their native country. Even at the end of his life he stated his belief 'that there is nothing more legitimate, more patriotic than to combat men who revolt against the law'.[44] Bugeaud for one was in no way influenced by the thinking of his adversaries, unlike his countryman Le Mière de Corvey, as we shall see presently. I was Bugeaud's, not Hoche's approach, which Callwell would later see as most appropriate for dealings with extra-European adversaries.[45]

Le Mière de Corvey

Jean Frédéric Auguste Le Mière de Corvey (1770–1832), musician, composer, and idealist, signed up to a Republican battalion in the Vendée out of Republican conviction. He took part in the Napoleonic Wars against Prussia, accompanied the Emperor to Poland, and for three years took part in the Peninsular War, fighting the Spanish insurgents. He only took his leave of the Grande Armée after its defeat at Waterloo. But even after that, he retained an interest in the use of 'irregular corps, light troops, and above all partisans commanded by a well-instructed and enterprising chief'.

Quite unlike Bugeaud, he developed a grudging respect in Spain for the 'bandes de partisans', his adversaries, poorly disciplined, poorly armed, and yet sincerely patriotic and capable of great feats. In the work he dedicated to them in 1823, he analysed their tactics, noting particularly their penchant for avoiding classical battles ('any engagements of soldiers of the line with our armies') and their harassment of small French detachments.[46] In his view, the Vendeans had been less successful than the Spaniards because they had organised themselves by parish and bishopric, while the Spaniards also coordinated their efforts at a national level. Le Mière extrapolated from these two insurgencies:

> In undertaking a national war in order to preserve one's independence and to defend against a foreign invasion, one has to undertake a war of extermination, and the enemy must needs be chased away, or else the nation that is on the defensive must be invaded, and the victor has the right to treat it as a conquered country.... If one wants to retain one's conquest, however, one has to treat the defeated with kindness, that is the only way to win their affection. Unfortunately all the wars in which we use massive levies [les levées en masse] are fed by some sort of fanaticism, be it a factional spirit, ideology [opinion], religion, etc. Without that there would not be wars of extermination the results of which are horrible.[47]

Le Mière tried to codify this type of people's resistance because 'one would render a great service to all countries which, after an unlucky war, find themselves defenceless before the aggression and the invasion by the conqueror.' He tried to prove 'to the victorious peoples that it is folly to want to invade a country when the threatened nation is predisposed to take such an imposing stance', i.e. 'to rise up en masse'. With this aim in mind, he proposed the creation of a 'central council', inspired by the Spanish juntas, to organise such a resistance.[48]

The partisan units Le Mière planned for were supposed to be made up of volunteers clad in simple uniforms.[49] He wanted to divide France up into large areas or 'circles' corresponding to local identities, which he still referred to as 'nations' (!), such as the Normans, the Picards, the Bretons, the Gascons, the Lorrainers, the Alsatians, and the Angevins, as local 'partisan-soldiers' would be more strongly motivated to defend their home region than mixed forces recruited from all over France.[50] Le Mière pointed to the Vendeans as a particularly good example of patriots committed to self-defence.[51] The 'partisan-soldiers' would not need very intensive training as they would not execute any complicated manoeuvres or marches. Instead, they would organise ambushes against enemy convoys, using 'stratagems and war ruises'.[52] He also envisaged suicide missions executed by the 'enfants perdus', an old term for soldiers spearheading assaults or otherwise sent on particularly dangerous but crucially important missions.[53] And finally, the siege of Zaragoza had made him ponder the possibility of a similar defence of Paris – at a time when it had not yet been cut open by Hausmann's *boulevards* – even though the walls surrounding the French capital were much less impressive than those of the Spanish city.[54]

Thus in applying the lessons of his own experiences in counter-insurgency operations to plans for the defence of his own country, Le Mière turned from an agent of repression into a disciple of the insurgents.

Brandt

Another witness of the Spanish Guerrilla was Heinrich von Brandt (1789–1868), who was born in Prussian-occupied Poland, near Poznan. Having studied law at Königsberg in East Prussia, he joined a Prussian *Freikorps*, or volunteer unit, in order to continue the Prussian fight against France when the Prussian government had ceased its resistance to Napoleon with the Peace of Tilsit of 9 July 1807. Brandt was an adventurer at heart, with the cause apparently of secondary importance to him. He went on to join the Polish Legion of the Vistula, which put itself at the service of Napoleon and was sent to Spain to fight the insurgency there.

Then in 1812, Brandt took part in Napoleon's Russian campaign on the side of the French forces. After the defeat of Napoleon, he was admitted to the Prussian officer corps thanks to the intervention of Gneisenau, becoming an instructor for officer cadets from 1819. A decade later, he became a teacher at the Prussian 'General War School' or military academy then directed by Clausewitz.[55] In 1823 he had published his memoirs on the Spanish Guerrilla from which he drew his own lessons for the conduct of irregular warfare. Thus he contributed to the very topical debate over whether an insurgency could succeed without external help or whether it needed a victory by regular forces elsewhere against the common enemy to achieve the crucial breakthrough. Brandt opined that the Spaniards had needed the help of the British to prevail over the French armed forces, a viewpoint which John T. Jones had already put forward in Britain.[56]

In 1829, Brandt published a field manual on the art of war, which contained a large section on people's war (*Volkskrieg*). Clausewitz and Brandt probably worked out their definition of the *Volkskrieg* or popular insurgency together. Given the contemporary adulation of the state, Brandt defined this *Volkskrieg* as a form of war in which all the energies of a people are put at the disposal of the head of state and where the majority of the population does not hesitate to sacrifice itself for the fatherland. Despite (or because of?) his colourful past, we find Brandt here defending a narrowly Prussian viewpoint.[57]

In practical terms, Brandt did not argue for an immediate insurgency to fend off an invasion. He counselled patience until such an invasion force had established itself (after the surrender of one's own regular forces) and felt safe, relying on long and vulnerable supply lines. Then the population should call upon pre-established rifle clubs (*Schützenvereine*). Brandt saw these as purely defensive. While admitting that the families of these snipers or militiamen were vulnerable to the occupying forces' reprisals, he saw no need in case of an occupation of the country to evacuate all the civilian population to remote and less accessible areas. These militiamen were to adopt small war tactics, themselves retreating into marshland, from there to launch small attacks on the occupying forces. Meanwhile, the population was to refuse all cooperation with the latter. Thus an insurgency would be carried out by a passive but hostile population on the one hand and an active minority of fighters, while waiting for regular forces to re-form to tackle the enemy's regular troops in a proper counter-attack. Inspired by the actions of the Spanish and the Russians against Napoleon, Brandt put his money on defeating the enemy far from the latter's home country, forcing him into a long and perilous retreat. In his view, 'a people's war which is not carried out with the support of a regular army and which moves outside its own frontiers would be risky and unnecessary. It would probably lead to the devastation of the country and yet would not have sufficient effects' on the final outcome of hostilities. He illustrated this with the Bavarian insurgency of 1705 against the Austrians, in the course of which 30,000 insurgents were defeated by the Austrian army, which, by way of reprisals, destroyed hundreds of villages.[58]

Alongside Clausewitz and Gneisenau, who would both die of cholera on this campaign, Brandt took part in the preventative deployment of Prussian units to Western Poland. Brandt thus had further occasion to study such uprisings close up. Yet in his later publications, he hardly mentioned what he had called people's war elsewhere, i.e. the national uprising that the Prussian crown feared. Prussian monarchs much preferred tightly controlled special forces to the nation in arms. Brandt, by now a staunch defender of Prussian interests, in 1848 again played his part in quashing a Polish insurgency, and won a battle against the Polish forces under Colonel Dombrowski. Thereafter Brandt made a short appearance in politics as a deputy at the German national assembly of Frankfurt upon Main in 1848, and then in the Prussian parliament of 1849–1850. He was then appointed commandant of the capital of his native region and finished his successful career in 1857 as a general.[59]

Polish exiles and the Polish insurgencies

As most of Poland was under Prussian and Russian occupation during the period considered here, a number of Poles who refused to come to an arrangement with this had sought refuge abroad, especially in France, since Napoleon had supported Polish ambitions to create a Polish nation-state. The works of these Polish exiles, many of whom had first-hand military experience, dwelt on the question whether regular forces would be needed to overthrow and chase away the occupying forces, or whether an insurgency of irregulars might suffice.[60]

Among them was Karol Bogumił Stolzman (1793–1854). At the age of 20 he had volunteered to join the forces of the Duchy of Warsaw. In 1813 he took part in the battles of Lützen, Bautzen, and Leipzig. As an officer of Congress Poland he was present at the creation of an artillery school for which he translated works from German. In 1830–1831 he took part in the Polish insurgency against Russia, after which he fled to France. It is only in 1844 that he published his book on *Partisan Warfare*.[61] Emanuel Halicz has noted that Stolzman recycled large sections from the writings of Mazzini. Stolzman was indeed in touch with the Italian Carbonari in Paris, which became a centre of refuge for exiles from various parts of Europe.[62]

In his book, Stolzman largely neglected the role that had been played by regular forces in the Peninsular War and in Russian resistance against Napoleon in 1812. He thought that the defeat of the 1830–1831 insurgency was above all due to Polish attempts to meet like with like. Instead, he advocated an alternative method:

> It is the same [method] which, more or less regulated, more or less energetically adopted, ensured victory to the Albanians under Skanderbeg, to the Serbs under Kara-Gieorg and to Miłosz [= Miloš Obrenović?] over the Turks and in our own time to the Greeks, and the Dutch over Philip II, to the Swiss over the house of Hapsburg and Charles of Burgundy, to America over Britain, to the Russians, Germany and Spain over the genius and forces of Napoleon.... This method is partisan warfare.[63]

In florid language, he thus advocated a large-scale insurgency: 'There is no act of treachery capable of crushing a partisan war; its fate cannot be decided by the enemy's occupation of the capital city. No single success on the battlefield can guarantee the enemy decisive victory.' For him the Spanish Guerrilla was the exemplary model, even though he had not experienced it himself.[64]

His compatriot Wojciech Adalbert Chrzanowski (1788/1793–1861) had also enlisted in the Army of the Duchy of Warsaw and had then taken part, alongside the French and against Russia, in the campaigns of 1812 and 1815. Subsequently, like Stolzman, he served Congress Poland as an officer, but then fought alongside Russian forces in the Russo-Turkish War of 1828–1829, and was promoted by the Tsar to the rank of lieutenant colonel. Nonetheless, again like Stolzman, he participated in the Polish anti-Russian uprising of the following year. He was even promoted to the rank of general by the insurgent leader Jan Krukowiecki, and was made governor of Warsaw.[65] When the Russians besieged Warsaw, he eventually surrendered in order to save the remaining civilian population from bloody reprisals; extremists held this against him and accused him of treason. Warsaw fell, and Chrzanowski was obliged to swear loyalty to Tsar Michael I. Chrzanowski then fled to Paris where, in 1835, he published his first book about partisan warfare.[66] Ostracised as a traitor by the other Polish exiles, he left Paris for Brussels and then for the New World, where he briefly settled in Louisiana. In 1849 he accepted the proposal of King Charles-Albert of Piedmont to lead his army against the Austrian forces commanded by the famous Marshal Radetzky, but the campaign ended with the defeat of the Piedmontese. In the following year, Chrzanowski left Turin for France where he died in 1861.[67]

His 1835 publication on partisan warfare – here as in Stolzman's writings used in the sense of insurgency – was translated into German and first published in Coblenz in 1839. From this he later took a large part that he put into a new book called *Little War Manual for Officers*.[68] In both works he claimed to be filling a void as there was so little literature on (ideologically driven) partisan warfare.[69] One of the distinguishing features of his writing was his insistence on good treatment of the soldiers, the need to gain and keep their trust, and the need for military hospitals.[70]

Unlike his compatriots, Chrzanowski openly admired the Russians for their achievements of 1812. He saw the role of the partisans as that of irregular forces fighting alongside regular forces, especially as long as the country had not yet been entirely overrun by the enemy. The main mission of partisans would be to fight behind enemy lines and on their flanks – much as the Russian irregulars had done in 1812. His observations led him to believe that partisans had to be trained and organised by the government in peacetime, on the model of local militias – here he was in implicit agreement with Brandt and Clausewitz. He assumed that, animated by patriotism and the sense of self-sacrifice, all young men would be ready to defend their country.[71] He had already understood that time favours the insurgent, and noted that the irregulars had to avoid direct confrontations with enemy regulars. Instead, they should attack lines of communication and small enemy detachments, seek to capture their officers, and force the enemy to go about only in large units (echoes of the Guerrilla). Thus one would demoralise an enemy army even if it was very numerous.[72]

Another Polish émigré in Paris addressed the same subject area in his 1835 publication *the System of Partisan Warfare as Discussed among the Émigré Community*.[73] This was Captain of artillery Wincenty Nieszokoć (1792–1865?), also a veteran of the campaigns of 1813–1814. His book was an explicit comparison of the Spanish Guerrilla with the Polish Insurgency of 1830–1831. Patriotically, he asserted that the Poles had shown themselves quite as committed as the Spanish. He pointed out that the latter had won because they had the British regular forces on their side, and: *'les gros bataillons ont toujours raison'*.[74] Going to the centre of the older debate about whether irregulars/insurgents can win a war on their own, he thus disagreed with Chrzanowski's interpretations and prescriptions, arguing that even 400 companies of irregulars could not match the 'unity of organisation which is the soul of a regular army', as they lacked its authority and cohesion.[75] Nieszokoć acknowledged that the strategy proposed by Chrzanowski and other Polish émigrés 'has been known from time immemorial' but had never met with success and had therefore 'never been adopted as a system of national defence'. He saw it as useful only as complementary to the regular operations of regular armed forces.[76] He thus sharply refuted the approach of Chrzanowski:

> Experience has convinced us how little advantage there is to be gained from partisan wars. They serve only to multiply disasters and the useless loss of blood, they have no effect on the course of the campaign as a whole and lead to the senseless waste of forces dispersed along the flank and rear of the enemy army which, if they were concentrated at its head, might lead to victory in a pitched battle. The harassment of

a strong enemy army along its flank and rear may be painful for it but it is never fatal. It has been correctly said before that '*les affaires ne se décident pas à la queue mais à la tête.*'[77]

Thus the Guerrilla served as both a positive and a negative template for this generation of Poles who had not themselves participated in this campaign. This also applied to Ignacy Prędzyński (1792–1850), yet another commander of the Polish Insurgency of 1830, who had accompanied Napoleon's forces during the campaign of 1812 and also at the battles of Leipzig and Waterloo. As one of the proponents of an unconditional defence of Warsaw against the Russians, he fell out with Chrzanowski in 1831 and became the strongest critic of his views.

Prędzyński distinguished between two sorts of 'small war' in a more sophisticated manner than most of his contemporaries. On the one hand, he applied the term to uprisings where population itself rises up to fight in small units (along the Spanish model) against the occupation forces. The other type of small war would see the mass of the population staying passive, but supporting irregular soldiers who harass the enemy behind his lines, on his flanks, his lines of communication, etc.[78] Prędzyński did not think his Polish compatriots capable of applying the first model. He showed considerable understanding of the Peninsular War, as confirmed by recent historiography[79]:

> Although no nation is as predisposed as the Spanish to banditry and guerrilla fighting – a result not so much of the geographical formation of the country as of their national character and the whole social situation in Spain – even the Spanish guerrillas could not maintain the independence of Spain on their own. This country had powerful and effective assistance from all the land and sea forces of Great Britain. Wellington's regular army did not for a moment cease to dominate the field and won brilliant victories. Spain had its own regular army and innumerable fortresses all of which staunchly defended themselves. But despite this, Spain came near to submitting to Napoleon . . . .
>
> [In Poland in 1831] events had to be settled in the final instance by the regular army but small-scale fighting could and should have doubled its effect and contributed in no uncertain terms to its successful outcome.

Comparing the Polish Insurgency with that of the Vendeans, Prędzyński thought it 'possible that the burden of the alien yoke will eventually dispose the Polish people to small-scale fighting, but in order for that to happen there must be a great change in national character'.[80] This pessimism about his compatriots was based more on his own experience than on any theoretical arguments.

For Prędzyński as for many other authors, the Vendée uprisings against the French Revolution and the Spanish Guerrilla continued to be the important points of reference, even where they had no personal experience of it and had experienced the Russian resistance against the Grande Armée instead. Chrzanowski was the exception to this rule. One might suspect that it was Russophobia more than anything else that led his compatriots to a different attitude.

Conclusions

The period we have examined thus saw several examples of cultural learning, concerning both how to stage and how to quell insurgencies. We have noted here only fairly famous cases of military writers whose published works illustrate this. There must be many less famous individuals whose thoughts on the subject were recorded only in letters or other unpublished documents; much more research deserves to be conducted in this area. Many other soldiers, perhaps illiterate or otherwise unaccustomed to recording their thoughts in any way, may well have been conscious of their experiences of the revolutionary wars they witnessed, and may in turn like Emmerich have borne witness with their lives to a revolutionary spirit, a revolution of the mind,[81] which aroused European peoples to stage their own uprisings for self-determination.

A revolutionary spirit as such was not necessarily 'translated' into a quest for democracy. Self-determination, the shaking off of a foreign yoke or foreign ruler, seems to have sufficed for most. Others seem to have been inspired by their experience of insurgencies only to embrace an all the more fervent determination to suppress them and root them out. Nor was the transition from small war as special force operations to small war in the sense of people's war complete: the term 'small', *kleiner Krieg, petite guerre,* even *la guerrilla* war continued to be used in both senses, and ideologically motivated insurgencies in the views of some authors were best fought by trained irregular units (parties) fighting alongside regular armies of patriots, with only the motivation distinguishing them from the way of war of the previous century. For Wilhelm Rüstow, perhaps the most important Swiss writer on the subject in the nineteenth century, '*der Kleine Krieg*' comprised both the actions of specialised irregular forces and, if necessary, local populations. It would always be a sideshow to the actions of the main forces defeating the enemy, unless the main forces had collapsed, and one only had the option of trying to exhaust him. Like the Poles, Rüstow and his Swiss compatriots could only see all of this in the context of the patriotic defence of their country against a foreign occupier.[82]

Back in 1810–1811, Clausewitz had lectured at the General War School on 'small war' in the sense of special operations.[83] In pursuing his reflections on strategy further, Brandt in the 1830s continued to use the term 'small war' in the following definition:

> Small war is conducted according to the general rules and principles of major [war], but with the constant modifications which the feeble strength of the troops used for this purpose necessitates. The smaller strength of these detachments allows a liveliness in their deployment which often contrasts vividly with the constrained and wooden forms of major war. Precisely in this way it becomes possible to realise the main principle of the small war – most extreme mobility and adaptability to all circumstances – and to find the necessary forms quickly for all nuances of this form of war: patrols, vanguards and operations of detachments. These are in no way constant.[84]

His book on 'small war' published in 1837 seemed to come back full circle to small war as the operations of special forces complementing those of regular armies. By contrast, as we have seen, Clausewitz and Brandt both decided to translate the Spanish term *guerrilla* not as 'small war' (*'Kleiner Krieg'*), but as *'Volkskrieg'*, 'people's war'.[85]

The paradigm shift was thus not complete. Moreover, as we have seen from the biographies of the authors discussed, almost all of them at some point served different countries successively – as mercenaries, one might argue. This did not mean, however, that none of them were committed patriots: in fact, curiously, some like the Poles and Clausewitz chose foreign employment precisely because they wanted to resist the power occupying their own country, indirectly, by joining an outside power also fighting that occupier. Distinctions between 'national' soldier and mercenary or between bandit and freedom fighter thus remained quite fluid, along with the terminology employed.

The common theme in the revolutionary spirit shared by most of the writers discussed in this article is one of a move towards greater independence and self-determination, which the return of the *anciens régimes* tried their best to contain, sometimes fighting like with like, insurgents with irregular crack troops to quell them. The fight for popular self-determination meant the shift from royal armies to people's armies, a greater value bestowed upon the soldiers themselves, and a refusal to submit to foreign rule, even if this meant insubordination and rebellion, and thus people's war.

To sum up: the period from 1775 to *c.*1830 saw a crucial transformation in the conceptualisation of small war, albeit one which did not give way to homogeneity of definition or understanding of the context, or preference for its tactics. Although the rest of the nineteenth century would see insurgencies and counter-insurgency campaigns of all sorts, as well as the use of special forces, and a small handful of important practitioners, it would produce little by way of theorising about the subject; what little there was, was conceptually not very original.[86] This lends all the more prominence to the previous period, where for the first time ideology and asymmetric warfare were linked not only in practice but also in theoretical conceptualisation. Indeed, we do see that between 1775 and 1830, theories and practice of how to wage and combat small wars spread from country to country, even across continents. Resulting variations there were many. Ideas of liberty and self-determination mattered hugely, firing reactionary defence of Catholicism and monarchy as much as nationalism and republican movements.

Notes

1. Clausewitz, *On War* (VIII.2), 580.
2. Quoted in Martin Rink's contribution to this special issue.
3. Berenhorst, *Betrachtungen*.
4. See my Introduction to this issue.
5. Guibert, *Journal*, 1: 39.

6. Paret, 'Revolutionary War and European Military Thought'; Paret, 'Colonial Experience'.
7. Selig and Skaggs, 'Introductory Essay'.
8. In the sense of leader of a *parti*, unit of special forces.
9. Starkey, *European and Native American Warfare*; Schmidt, 'Der Guerrillero', 169.
10. Quoted in Weighley, *American Way of War*, 36.
11. Emmerich, *Partisan's War*. For the former standard works, see Grandmaison, *La Petite Guerre*; Jeney, *Le Partisan*.
12. Strieder, *Grundlagen*, quoted in Siebert, *Hanauer Biographien*, 435; Heitzer, *Insurrectionen*, 140–5.
13. Ewald, *Den kleinen Krieg*; *Gedanken* (Anon.); *Dienst der leichten Truppe*; *Belehrungen über den Krieg*; *Folge der Belehrungen über den Krieg*.
14. Ewald, *Den kleinen Krieg*; *Dienst der leichten Truppe*.
15. Ewald (Anon.), *Gedanken*, 78.
16. Ewald, *Den kleinen Krieg*, preface, 5.
17. Klipstein, *Theorie des Dienstes*; Valentini, *Abhandlung*.
18. Thiele, *Gneisenau*; see also Martin Rink's contribution to this issue.
19. Falkenstein, *Thaddäus Kosciusko*.
20. Anon., *Czy Polacy*, cited by Halicz, *Partisan Warfare*, 37.
21. Novak, *History and Geopolitics*, 119.
22. Joes, 'Insurgency and Genocide', 29.
23. Bonnet, *Guerres insurrectionnelles*, 132.
24. Ibid., 132f.
25. Ibid., 133s.
26. Callwell, *Small Wars*, 71–107.
27. L[ilienstern], *Handbuch für den Offizier*, 2: 58; Clausewitz, *On War*, 188, 281.
28. Archduke Charles, 'Geist des Kriegswesens überhaupt', *c*.1823–1826, in Waldtstätten, *Erzherzog Karl*, S.110f.
29. Quoted in Bois, *Bugeaud*, 83f.
30. On the reality behind this myth, see Charles Esdaile's contribution in this issue.
31. See Martin Rink's contribution in this issue.
32. Jomini, *L'Art de la Guerre*, 1: 72, 77s.
33. Guibert, 'Essay général de tactique', in Guibert, *Stratégiques*, 137f.
34. Clausewitz, *On War*, 592.
35. In 'Bekenntnisdenkschrift' (1812), see Rothfels, *Clausewitz*, 85–8.
36. In Clausewitz, *Schriften – Aufsätze – Studien – Briefe*, 1: 321, 'General Wassington' [*sic*].
37. See Kaempff, 'Lost through Non-Translation'.
38. Clausewitz, *On War*, VI.26.
39. Bois, *Bugeaud*, 76f.
40. Quoted in Ibid.
41. Iberian town that rose up against the Romans. When the Romans defeated the town, a part of the population preferred to commit suicide rather than surrendering. Numantia became a rallying myth for the Spanish Guerrilla, as this letter of Bugeaud demonstrates.
42. Quoted by Bois, *Bugeaud*, 80.
43. Bugeaud, *De la stratégie*, reprinted in Bugeaud, *Par l'épée*, see 110–18.
44. Bugeaud, *La Guerre des Rues*, 109.
45. Callwell, *Small Wars*, 146.
46. Le Mière de Corvey, *Des Partisans*, iiif.
47. Ibid., vii–ix.
48. Ibid., xiv s.

49. Ibid., xvi.
50. Ibid., 108–11.
51. Ibid., 144–6.
52. Ibid., 150–98 *passim*.
53. Ibid., 153–6; 199–205.
54. Ibid., 219–30.
55. See Priesdorff, *Soldatisches Führertum*, 6, 575.
56. Brandt, *Ueber Spanien*, 57–75; Jones, *War in Spain*.
57. Brandt, *Handbuch für den ersten Unterricht*.
58. Ibid., 321–363.
59. Priesdorff, *Soldatisches Führertum*; Killy and Vierhaus, *Deutsche Biographische Enzyklopädie*, 2: col. 423.
60. Halicz, *Partisan Warfare*; Beckett, *Modern Insurgencies*, 15.
61. Stolzman, *Partyzantka*.
62. Halicz, *Partisan Warfare*, 48.
63. Quoted in ibid., 82.
64. Ibid.
65. Bolek, *Who's Who in Polish America*, 431.
66. Chrzanowski, *O wojnie partyzanckiej*.
67. Hoefer, *Nouvelle Biographie générale*, 10, 117; Wurzbach, *Biographisches Lexikon*, 2: 112.
68. Général C***, *Kleines Kriegshandbuch*.
69. Chrzanowski, *Ueber den Partheigaenger-Krieg*.
70. Ibid., 23–7, 32s.
71. Ibid., 5f., 27.
72. Ibid., 10–12.
73. Nieszokoć, *O systemie wojny*.
74. In French in the original: 'the big battalions are always right'.
75. Nieszokoć, *O systemie wojny*, quoted in Halicz, *Partisan Warfare*, 73.
76. Quoted by Halicz, *Partisan Warfare*, 73f.
77. Quoted ibid., 47; French in the original: 'things are not decided at the back but at the front'.
78. Quoted ibid., 22.
79. Esdaile, *Fighting Napoleon*.
80. Quoted Halicz, *Partisan Warfare*, 23.
81. Hobsbawm, *Nations and Nationalism*; Israel, *Revolution of the Mind*.
82. Rüstow, *Lehre vom Kleinen Kriege*. The Swiss case is of course particularly interesting as Switzerland has a long tradition of homeland defence reaching back to the Middle Ages, and even in the twentieth century prepared for a 'people's war' against any invader. See Mantovani, 'Der "Volksaufstand"'.
83. *Carl von Clausewitz: Schriften - Aufsätze - Studien – Briefe*, Vol. 1.
84. Brandt, *Der Kleine Krieg*, 2.
85. Ibid.
86. With the exception, perhaps, of Auguste Blanqui's writings on urban guerrilla.

Bibliography

Primary sources

Amblard, André. 'Les guerres de Vendée vues par un ardéchois républicain, 1793–1999 (Le journal du capitaine Amblard, de Lussas (Ardèche)'. *Revue de la Société des Enfants et Amis de Villeneuve-de-Berg* 41 (1981): 50–70.

Amiot Père Joseph-Maire, and Joseph de Guignes, trans. and ed. *Art militaire des Chinois, ou Recueil d'anciens traités sur la guerre, composés avant l'ere chrétienne, par différents généraux chinois.* Paris: Didot l'ainé, 1772, containing 'Les Treize articles sur l'art militaire, ouvrage composé en chinois par Suntse . . .'.

Anon. ('Le petit diable boiteux'). *Mes réminiscences de l'Espagne: . . . tactique des guérillas et des miquelets . . .* Paris: Constant-Chantipie, 1823.

Anon. (a Prussian officer). *Abhandlung über den kleinen Krieg und über den Gebrauch der leichten Truppen, mit Rücksicht auf den französischen Krieg* mit Anmerkungen von L.S. von Brenkenhoff. Berlin: Christian Friedrich Himburg, 1799.

Anon. *Czy Polacy wybićsięmogę na niepodległość ? [Can the Poles achieve their independence?].* 1st ed. 1800; reprinted several times until 1831.

Anon. *Galien rethoré.* Paris: Antoine Vérard, 1500, with an edition in modern French as *Galien le restoré en prose.* Edited by H.-Er. Keller and N.L. Kaltenbach. Paris: Champion, 1998.

Anon. *Instruccion de guerrilla.* Spoleto: Tipografia Bossi & Bassoni, 1849.

Anon. [Fourquevaux, Raymond.] *Instruction sur le fait de la guerre.* Paris: Michel Vascosan, 1548. For a translation of excerpts into English, see Beatrice Heuser ed. *The Strategy Makers*, 32–49.

Arndt, Ernst Moritz. *Katechismus für den teutschen Kriegs= und Wehrmann.* Leipzig: Rein, 1813.

Aschmann, Birgit. *Preußens Ruhm und Deutschlands Ehre. Zum nationalen Ehrdiskurs im Vorfeld der preußisch-französischen Kriege des 19. Jahrhunderts.* München: Oldenbourg, 2013.

Artola-Gallego, Miguel, ed. *Memorias del general don Francisco Espoz y Mina.* 2 vols. Madrid: Ediciones Atlas, 1962.

Aubigné, Antoine d'. 'Lettre à M. l'evesque de Maillezais'. In *Œuvres complètes de Théodore Agrippa d'Aubigné*, Vol. 1. Edited by Eugène Réaumé et François de Caussade. Paris: A. Lemerre, 1873.

Aulard, Alphonse, ed. *Recueil des Actes du Comité de Salut Public avec la correspondance officielle des représentants en mission.* Paris: Imprimerie Nationale, 1889–1895.

Bald, Detlef. 'Wehrpflicht-Der Mythos vom legitimen Kind der Demokratie'. In *Die Wehrpflicht. Entstehung, Erscheinungsformen und politisch-militärische Wirkung*, edited by Roland G. Foerster, 30–45. Munich: Oldenbourg, 1994.

Ballesteros, Francisco. *Respetuosos descargos que el Teniente General Don Francisco Ballesteros ofrece a la generosa nación española en contestación a los cargos que S.A. la Regencia del Reino se ha servido hacerle en su manifiesto del 12 de diciembre del año pasado de 1812 dirigido a la misma para su inteligencia.* Cádiz, 1813. Instituto de Historia y Cultura Militar, Colección Documental del Fraile, vol. 154.

Balvay, Arnaud. *L'épée et la plume: Amérindiens et soldats des troupes de la marine en Louisiane et au pays d'en haut (1683–1763)*. Laval: Presses de l'Université de Laval, 2006.

Balzac, Honoré de. *Les chouans*. 3rd edition, Paris: Librairie de Werdet, 1836.

Beckett, Ian F.W. *Modern Insurgencies and Counter-Insurgencies: Guerrillas and their Opponents since 1750*. London: Routledge, 2001.

Berenhorst, Georg Heinrich von. *Betrachtungen über die Kriegskunst, über ihre Fortschritte, ihre Widersprüche und ihre Zuverlässigkeit*. 1st ed. 1797; 2nd ed. Leipzig, 1798; 3rd ed. 1827 reprinted Osnabrück: Biblio Verlag, 1978.

Berlioz, Jacques and Marie-Anne Polo de Beaulieu, eds. *Les exempla médiévaux: Introduction à la recherche suivie des Tables critiques de l'Index exemplorum de F.C. Tubach*. Carcassonne: GARAE-Hésiode, 1992.

Berlioz, Jacques and Marie-Anne Polo de Beaulieu, eds. *Les exempla médiévaux: nouvelles perspectives*. Paris: H. Champion, 1998.

Berlioz, Jacques and Marie-Anne Polo de Beaulieu. 'Les prologues des recueils d'*exempla*'. In *Les prologues médiévaux*, edited by Jacqueline Hamesse, 275–320. Turnhout: Brepols, 2000.

Bianco, Carlo and Conte di Saint Jorioz. *Della guerra nazionale d'insurrezione per bande*. 1st ed. 1830; reprinted Rome: Robin, 2011.

Blayney, Andrew. *Narrative of a Forced Journey through Spain and France as a Prisoner of War in the Years 1810 to 1814*. London: E. Kirby, 1814.

Bonaparte, Napoleon. *Correspondance de Napoléon 1er publiée par l'ordre de l'Empereur Napoléon III* [*Correspondence of Napoleon* printed by order of the Emperor Napoleon III]. 32 vols. Paris: Imprimerie Impériale, 1858–1869.

Boyen, Hermann von. *Erinnerungen aus dem Leben des Generalfeldmarschalls Hermann von Boyen*. edited by Dorothea Schmidt. Berlin (East): Brandenburgisches Verlagshaus, 1990.

Brandt, Heinrich von. *Handbuch für den ersten Unterricht in der höhern Kriegskunst*. 1st ed. Berlin, 1829; reprinted in Wiesbaden: LTR Verlag, 1981.

Brandt, Heinrich von. *Der Kleine Krieg in seinen verschiedenen Beziehungen*. 1st ed. 1837; 2nd ed. Berlin: Friedrich August Herbig, 1850.

Brandt, Heinrich von. *Ueber Spanien mit besonderer Hinsicht auf einen etwaigen Krieg*. Berlin: Schüppel'sche Buchhandlung, 1823.

Bueil, Jean de. *Le Jouuencel*. Paris: Antoine Vérard, 1493. Edited by Guillaume Tringant, Camille Favre, and Léon Lecestre. Paris: H. Laurens, 1887.

Bugeaud de la Piconnerie, Duc d'Isly. *Part la Charrue Ecrits et Discours*. Edited: by Paul Azan. Paris: PUF, 1948.

Bugeaud, de la Piconnerie, Duc d'Isly, and Thomas Robert. *Aperçus sur quelques détails de la guerre*. 3rd ed. Paris: Leneuve & Dumaine, 1846.

Bugeaud de la Piconnerie, Duc d'Isly and Thomas Robert. *De la stratégie, de la tactique, des retraites et du passage des défilés dans les montagnes des Kabyles*. Alger: Imprimerie du Gouvernement, 1850.

Bugeaud, de la Piconnerie, Duc d'Isly, and Thomas Robert. *La Guerre des Rues et des Maisons*. MS of 1849; printed Paris: Jean-Paul Rocher, 1997.

Bugeaud, de la Piconnerie, Duc d'Isly, and Thomas Robert. *Œuvres militaires*. Paris L. Baudoin, 1883; reprinted 1982.

Bunbury, Thomas. *Reminiscences of a Veteran, being Personal and Military Adventures in Portugal, Spain, France, Malta, New South Wales, Norfolk Island, New Zealand, Andaman Islands and India*. London: Charles J. Skeet, 1861.

Bystrzonowski, Louis [Ludwik] de Szafraniec. *Notice sur le Réseau Stratégique de la Pologne pour Servir à une Guerre de Partisans*. Paris: Bourgogne et Martinet, 1842.

C***, General [Wojciech Chrzanowski]. *Kleines Kriegshandbucvh für Offiziere: Abriss der angewandten Taktik aller Waffen, der Generalstab und der Parteigängerkrieg.* Halle: C.A. Schwetschke & Sohn, 1852.

Cabello, F., F. Santa Cruz, and R.M. Temprado. *Historia de la guerra última en Aragón y Valencia.* 2 vols. Madrid: Imprenta del Colegio de Sordo-Mudos, 1845.

Callwell, C.E. *Small Wars: Their Principles and Practices.* 3rd ed. London: HMSO, 1906.

Carvajal, José Maria. *Reglamento para las Partidas de Guerrilla.* Cádiz: Don Nicolas Gomez de Requena, 1812.

Charles, Archduke. 'Das Kriegswesen in Folge der französischen Revolutionskriege'. (1838). In *Erzherzog Karl – Ausgewählte militärische Schriften,* edited by Freiherr von Waldtstätten, 201–216. Berlin: Richard Wilhelmi, 1882.

Charles, Archduke. 'Das Kriegswesen in Folge der französischen Revolutionskriege'. (1838). In *Erzherzog Karl – Ausgewählte militärische Schriften,* edited by Freiherr von Waldtstätten, 199–216. Berlin: Richard Wilhelmi, 1882.

Chrzanowski, Wojciech, *O wojnie partyzanckiej.* Paris: 1835. Translated by a Prussian Officer as *Über den Partheigängerkrieg: eine Skizze.* 1st ed. 1839, 2nd ed. Berlin: Stuhrsche Buchhandlung, 1846.

Chuikevich, Petr Andreevich. *Reflections on the War of 1812.* Translated from Russian by Alexis Eustafieve Boston: Munroe & Francis, 1813.

Clausewitz, Carl von. 'Lectures on Small War (1810–11)'. In *Carl von Clausewitz: Schriften – Aufsätze – Studien – Briefe,* Vol. 1. Edited by Hahlweg Werner. Göttingen: Vandenhoeck & Ruprecht, 1966.

Clausewitz, Carl von. 'Bekenntnisschrift von 1812'. In *Schriften–Aufsätze–Studien–Briefe,* vol. 1, edited by Werner Hahlweg, 678–751. Göttingen: Vandenhoek & Ruprecht, 1966.

Clausewitz, Carl von. *Politische Schriften und Briefe.* edited by Hans Rothfels. Munich: Drei Masken Verlag, 1922.

Clausewitz, Carl von. *Vom Kriege.* Berlin: Dümmler, 1932, translated as *On War.* Translated and edited by Peter Paret and Michael Howard. Princeton, NJ: Princeton University Press, 1976.

[Cölln, Friedrich von]. 'Briefe eines Reisenden von Berlin nach Königsberg in Preußen, im September 1807'. *Neue Feuerbrände* 1808, No. 10, 91–124. http://www.ub.uni-biel efeld.de/diglib/aufkl/nfeuerbraende/nfeuerbraende.htm

Commynes, Philippe de. *Chronique et histoyre [...] durant le regne du Roy Louys unziesme.* Paris: Pour François Regnault, 1529.

Conversations-Lexikon. 2nd ed. Leipzig: Brockhaus, 1817.

Córdova y Valcarcel, Fernando Fernández de. *Mis memorias íntimas.* 2 vols. Madrid: Ediciones Atlas, 1996.

Corneille, Thomas. *Le Dictionnaire des arts et des sciences.* Paris: Vve de J.B. Coignard, 1964.

Courtilz, Sandras de. *Histoire de la guerre de Hollande.* The Hague: Henri van Bulderen, 1689.

Covarrubias, Sebastián de. *Tesoro de la Lengua Castellana o Española,* 1611 ed. with additions from 1674 ed. edited by Martín Riquer. Reprinted Barcelona: S.A. Horta, 1943.

Dangeau, marquis de. *Journal du marquis de Dangeau, avec les additions inédites du duc de Saint-Simon.* Edited by E. Soulié and L. Dussieux. Paris: F. Didot, 1854–1860.

Davidov, or Davidoff, Denis Vasilevich. *Essai sur la guerre de partisans.* Translated from Russian by Count Héraclius de Polignac. Paris: J. Corréard, 1841.

De la, Croix and Armand François. *Traité de la petite guerre pour les compagnies franches.* Paris: Antoine Boudet, 1752.

Debidour, Antonin. *Recueil des actes du Directoire exécutif.* 4 vols. Paris: Imprimerie nationale/Leroux, 1910–1917.

Decker, Carl von. *Algerien und die dortige Kriegführung.* 2 vols. Berlin: Friedrich August Herbig, 1844.

Decker, Carl von. *Der kleine Krieg im Geist der neueren Kriegsführung oder Abhandlung über die Verwendung und den Gebrauch aller drei Waffen im kleinen Kriege.* 1st ed. 1822; 3rd ed. Berlin: Ernst Siegfried Mittler, 1828.

Delorme, Simon. *Journal de Simon Delorme à l'armée de Bavière en 1742.* edited by Claude Gaurat. Ph.D. Paris IV-Sorbonne [2007].

Dictionnaire de l'Académie française. Paris: Jean-Baptiste Coignard, 1694.

Diderot, Denis & d'Alembert. *L'Encyclopédie,* Paris: Briasson, 1751–1780.

Emmerich, Andreas. *The Partisan in War or the use of a corps of Light Troops to an Army.* London: H. Reynell, 1789. Translated into German by J.G. Hoyer, *Der Partheygänger im Kriege, oder Gebrauch der leichten Truppen im Felde.* Dresden: 1791. http://www.britishbrigade.org.library/emm1.html

Estienne, Robert. *Dictionnaire François latin.* Paris: Robert Estienne, 1549.

Ewald, Johann von, *Abhandlung über den kleinen Krieg.* Cassel: Johann Jacob Cramer, 1785. Translated into English by Robert A Selig and David Curtis Skaggs, *Treatise on Partisan Warfare.* New York: Greenwood Press, 1991.

Ewald, Johann von. *Abhandlung von dem Dienst der leichten Truppe.* Flensburg: Schleswig, and Leipzig, 1790 & 1796.

Ewald, Johann von. *Folge der Belehrungen über den Krieg, besonders über den kleinen Krieg, durch Beispiele großer Helden und kluger und tapferer Männer.* Schleswig: J.G. Röhß, 1800.

Ewald, Johann von. *Belehrungen über den Krieg, besonders über den kleinen Krieg, durch Beispiele großer Helden und kluger und tapferer Männer.* Schleswig: J.G. Röhß, 1798.

Ewald, Johann von (Anon.). *Gedanken eines Hessischen Officiers über das, was man bey Führung eines Detaschements im Felde zu thun hat.* Cassel: Johann Jacob Cramer, 1774.

Falkenstein, Karl. *Thaddäus Kosciusko: Nach seinem öffentlichen und häuslichen Leben geschildert.* 1st ed. 1827; 2nd ed. Leipzig: F.A. Brockhaus, 1834.

Fernández de Córdova, Fernando. *Tactica de guerrilla.* Madrid: s.e., 1870.

Flavius Vegetius Renatus. *Flave Vegece Rene [. . .], du fait de guerre et fleur de chevalerie [. . .]. Sexte Jules Frontin, [. . .], des stratagèmes, especes, et subtilitez de guerre [. . .].* Paris, C. Wechel, 1536.

Folard, Jean Charles Chevalier de: *Nouvelles découvertes sur la guerre, dans une dissertation sur Polybe* (Paris: Jean-François Josse & Claude Labottière, 1724).

Frederick II of Prussia. *Des Königs von Preussen Majestät Unterricht von der Kriegskunst an seine Generals.* Frankfurt & Leipzig, s.n., 1761.

Furetière, Antoine de. *Dictionnaire Universel.* La Haye: Arnout et Reinier Leers, 1690.

G[entilini], E[nrico]. *Guida del Milite.* Capolago: Tipografia e Libreria Elvetica, 1835.

Gentilini, Enrico. *Instruzione agli Italiani per la Guerra dell'Indipendenza Nazionale ad uso della Guardia Civica e de volontari stacorridori.* Dijon: Da Torchi Loireau-Feuchot, 1848.

Grandmaison, [Capitaine Thomas-Antoine le Roy] v. *La Petite Guerre, ou traité du service des troupes légères en campagne.* S.l., s.e., 1756.

Goethe, Johann Wolfgang von. *Kampagne in Frankreich.* Stuttgart: Karl Göpel, 1850.

Grand Corpus des dictionnaires [9e-20e s.]. Paris: Classique Garnier. http://www.classiquegarnier.com

Guibert, Comte de. *Stratégiques.* Paris: L'Herne, 1977.

Guibert, G.A.H., [sic!]. *Journal d'un Voyage en Allemagne fait en 1773*, 2 vols. Paris: Treuttel et Würtz, 1803.

Hamilton, Alexander. 'Hamilton to Washington'. In *Chapter in The Works of Alexander Hamilton*, Vol. 4. Edited by Henry Cabot Lodge. New York. Putnam, 1904. http://oll.libertyfund.org/title/1381/64393

Henegan, Richard. *Seven Years' Campaigning in the Peninsula and the Netherlands*. 2 vols. London: Henry Colburn, 1846.

Heuser, Beatrice, ed. *The Strategy Makers: Thoughts on War and Society from Machiavelli to Clausewitz*. Santa Barbara, CA: ABC Clio for Praeger, 2010.

Hoche, Lazare. *Authentic copies of the instructions given by Gen[eral] Hoche to Colonel Tate previous to his landing on the coast of South Wales, in the beginning of 1797*. London: Wright, 1798.

House of Commons. *Report of the Committee of the House of Commons in consequence of the several motions relative to the treatment of prisoners of war. Including the whole of the examinations taken before the Committee the correspondence relative to the exchange of prisoners; the instructions of Colonel Tate, &c*. London: Wright, 1798.

Jacob, William. *Travels in the South of Spain in Letters written A.D. 1809 and 1810*. London: J. Johnson and Co., 1811.

Jeney, Capitaine de. *Le partisan, ou l'art de faire la petite-guerre avec succes selon le génie de nos jours*. The Hague: 1759. Translated by an officer in the Army [Roger Stevenson] as *The Partisan, or, the Art of Making War in Detachments*. London: R. Griffiths, 1760.

Jomini, Henri de. *Précis de l'Art de la Guerre, ou Nouveau Tableau Analytique*. Paris: Asselin, 1838.

Jones, John T. *Account of the War in Spain and Portugal, and in the South of France, from 1808 to 1814, inclusive*. 2 vols. London: T. Egerton, 1821.

Kircheisen, Friedrich, ed. *Memoiren aus dem spanischen Freiheitskampfe 1808–1811*, with texts by Ludwig von Grolmann, Albert Jean Michel Rocca, Moyle Sherer, Heinrich von Brandt, Henri Ducor, Don Juan Andrés Nieto Samaniego Hamburg: Gutenberg-Verlag, 1908.

Klipstein, Friedrich Leopold. *Versuch einer Theorie des Dienstes der leichten Truppen besonders in Bezug auf leichte Infanterie*. Darmstadt: G. F. Heyer, 1799.

la Roche, Conte. *Essai sur la petite guerre; ou méthode de diriger les différentes opérations d'un corps de 2500 hommes de troupes légères*. Paris: s.e., 1770.

La Curne Sainte Palaye, Jean-Baptiste de. *Dictionnaire historique de l'ancien langage François*, edited by L. Favre. Niort: L. Favre and Paris: H. Champion, 1875–1882.

La Noue, François de. *Discours politiques et militaires du seigneur de La Noue*. Basle: François Forest, 1587.

La Roche Aymon, [Charles Antoine Étienne Paul de]. *Ueber den Dienst der leichten Truppen*. Königsberg: Degen, 1808.

Larmessin, Nicolas de. *Les grandes victories du très puissant monarque Louis le Grand emporté sur ses ennemis par mer et par terre*. Paris, 1690.

Lavaux, François. *Mémoires de campagne, 1793–1814*. Edited by Cristophe Bourachot. Paris: Arléa, 2004.

Lawrence, Thomas Edward. *Seven Pillars of Wisdom*. 1st ed. 1922; London: Vintage, 2008.

Lyautey, Marshal Louis Hubert Gonzalve. *Rayonnement*. Writings edited by Patrick Heidsieck. Montreal: Gallimard/Editions Variétés, 1944; includes Louis-Hubert Lyautey, 'Du Rôle social de l'officier' originally published in *Revue des deux Mondes* (15 March 1891).

Ménage, Gilles. *Les Origines de la langue françoise*. Paris: Jean Anisson, 1694.

Mière, de Corvey and J.F.A. Le. *Des Partisans et des corps irréguliers*. Paris: Anselin & Pochard et al., 1823.

Mieroslawski, Ludwig. *Kritische Darstellung des Feldzuges vom Jahre 1831 und hieraus abgeleitete Regeln für Nationalkriege aus dem polnischen von R.v.K.* 2 vols. Berlin: Behr, 1848.

Mieroslawski, Ludwik, *Mémoire justificatif dans le Débat entre l'organisateur général des forces polonaises et ses adversaires*. Paris, s.n., 1864.

Montecuccoli, Raimondo. *Memorie*. Köln: Henricus de Huyssen, 1704.

Montluc, Blaise de. *Commentaires de Blaise de Montluc*. Edited by Claude Bernard Petitot. Paris: Foucault, 1821–1822.

Nicot, Jean. *Thresor de la langue francoyse*. Paris: David Douceur, 1606.

Nieszokoć, Wincenty, *O systemie wojny partyzanckiej wzniesionym wśród emigracij*. Paris: s.e, 1835.

Ofarrill, Don Gonzales. *Instruccion que deben seguir los oficiales y tropas del 1er batallon de voluntarios de Cataluña quando se empleen en guerrilla, o como tiradores*. Liorna: Imprenta de Antonio Vignozzi, 1806.

Richelet, Pierre. *Dictionnaire françois*. Genève: J.-H. Widerhold, 1680.

Rohan, Henri duc de. *Le parfait capitaine*. Paris: s.n, 1636, n.p.

Romero Alpuente, Juan. *El grito de la razón al español invencible, ó La guerra espantosa al perfido Bonaparte de un togado aragonés con la pluma: discursos*. Zaragoza, 1808.

Romero Alpuente, Juan. *Wellington en Cádiz y Ballesteros en Ceuta*. Cádiz, 1813.

Rotteck, Karl von. *Ueber Stehende Heere und Nationalmiliz*. Freiburg: 1816.

R[ühle] v[on] L[ilienstern], [Otto August]. *Handbuch für den Offizier zur Belehrung im Frieden und zum Gebrauch im Felde*. Part 2. Berlin: G. Reimer, 1818.

Rüstow, Wilhelm. *Die Lehre vom Kleinen Kriege*. Zurich: Friedrich Schultheß, 1864.

Saint-Genies, Ray de. *L'officier partisan, suivi des stratagemes de guerre des francois*. 6 vols. Paris: de Lalain, Crapard, 1763–1769.

Santa Cruz, de Marcenado, Don Alvaro de Navia Osorio y Vigil, and Marques de *Reflexiones Militares* 11 vols. Vols. 1–10, Turin: Juan Francisco Mayrese, 1724–1727; vol. 11, Paris: Simon Langlois Vizconde de Puerto. 1730.

Saint-Simon, Claude-Henri de Rouvroz de. *Mémoires*, Vol. 1. Edited by A. Boislisle. Paris: Hachette, 1879.

Saurin de la Iglesia, María Rosa, ed. *Manuel Pardo de Andrade: Semanario Político, Histórico y Literario de la Coruña* (1809–1810). Vols. 2. Edición facsimile. La Coruña: Fundación Pedro Barrie de la Maza, 1991.

Savary, Jean Julien Michel. *Guerre des Vendéens et des Chouans contre la République française, ou Annales des départements de l'Ouest, par un officier supérieur des armées de la République habitant dans la Vendée avant les troubles*. Vols 6. Paris: Baudouin, 1824–1827.

Saxe, Maurice de. *Traité des légions*. The Hague: aux dépens de la compagnie, 1753.

Scharnhorst, Gerhard von. *Private und dienstliche Schriften*. Edited by Johannes Kunisch, Michael Sikora, and Tilman Stieve. Cologne: Böhlau, 2009.

Scharnhorst, Gerhard. *Handbuch für Offiziere*. Hannover: Helwing'sche Verlagsbuchhandlung, 1787–1790.

Scharnhorst, Gerhard, *Militairisches Taschenbuch zum Gebrauch im Felde*, 3rd ed. Hannover 1794; reprinted in Osnabrück: Biblio, 1980.

Schels, Johann Baptist. *Leichte Truppen; kleiner Krieg*. Vienna: Strauss, 1813.

Serrano Valdenebro, Jose. *Manifiesto de los servicios hechos a la patria por el Jefe de Escuadra, Don Jose Serrano Valdenebro, desde el movimiento de la nación a la justa defensa contra la invasión que nos tiraniza hasta su remoción del mando de la Sierra Meridional en 2 de julio de 1811*. Algeciras, 1811.

Stein, Freiherr vom: *Briefe und amtliche Schriften*, edited by Erich Botzenhartand Walther Hubatsch. Stuttgart: Kohlhammer, 1960.

Stolzman, Karol Bogumił. *Partyzantka czyli wojna dla ludów powstających najwłasciwśza* [Partisan warfare as the most appropriate for insurgent nations]. Paris & Leipzig: Brockhaus & Avenarius, 1844.

Strieder, Friedrich Wilhelm. *Grundlagen zu einer hessischen Gelehrten-und Schriftsteller-Geschichte*, 20 vols. Kassel, 1780–1863.

Stuart, Berault. Seigneur d'Aubigny. *Traité sur l'Art de la Guerre (1508)*. Edited by Elie de Comminges. Den Haag: Martinus Nijnhoff, 1976.

Thucydides. *Peloponnesian War*, here the French translation: *Lhistoire de Thucydude Athenien, de la guerre, qui fut entre les Peloponnesiens et Atheniens*. Paris: Josse Bade, 1527.

Tone, Theobald Wolfe. *The Writings of Theobald Wolfe Tone 1763–1798*. In *America, France and Bantry Bay August 1795 to December 1796*, edited by Theodore Moody, Robert McDowell, and Christopher Woods. Vol. 2. Oxford: Clarendon Press, 2001.

Turreau, Louis-Marie. *Mémoires pour servir à l'histoire de la Guerre de la Vendée*. Paris: Baudouin frères, 1824.

Valentini, Georg Wilhelm Freiherr von. *Abhandlung über den kleinen Krieg und über den Gebrauch der leichten Truppen mit Rücksicht auf den französischen Krieg*. 1st ed. Berlin: Christian Friedrich Himburg: 1799. 4th ed. Berlin: J. W. Boicke, 1820.

Ville, Antoine de. *De la Charge des Govvernevrs des Places*, 'latest ed.' Paris: s.n., 1640.

Villars, Maréchal de. *Mémoires du Marechal de Villars*, edition de Vogüe, 2 vols. Paris: 1887.

Voß, Julius von. *Der kleine Krieg*. Berlin: Schmidt, 1809.

Waldtstätten, Freiherr von, ed. *Erzherzog Karl – Ausgewählte militärische Schriften*. Berlin: Richard Wilhelmi, 1882.

Walter, Dierk. *Preußische Heeresreformen 1807–1870. Militärische Innovation und der Mythos der Roonschen Reform*. Paderborn: Ferdinand Schöningh, 2003.

Wuest, Baron de. *L'Art Militaire du Partisan*. La Haye: s.n., 1768.

Wurzbach, Constant von. *Biographisches Lexikon des Kaiserthums Österreich*. Vienna: Zamarski, 1857.

Zedler, Johann Heinrich, ed. *Grosses vollständiges Universal Lexicon*, Vol. 26. Leipzig and Halle: Joh. Hein. Zedler, 1740.

Secondary literature

Books

Alexander, Don W. *Rod of Iron: Counterinsurgency Policy in Aragon during the Peninsular War*. Wilmington, DE: Scholarly Resources, 1985.

Arfaioli, Maurizio. *The Black Bands of Giovanni: Infantry and Diplomacy During the Italian Wars (1526–1528)*. Pisa: Edizioni Plus, 2005.

Aróstegui, Julio, Canal, Jordi, Calleja, Eduardo. *Las guerras carlistas: hechos, hombres e ideas*. Madrid: La Esfera de los Libros, 2003.

Artola-Gallego, Miguel. *Los afrancesados*. Madrid: Ediciones Turner, 1976.

Artola-Gallego, Miguel. *La burguesía revolucionaria. 1808–1874*. Madrid: Alianza, 1974.

Artola-Gallego, Miguel. *Historia de la España contemporánea*. 2 vols. Madrid: Alleanza editorial, 1976.

Artola-Gallego, Miguel. *Los orígenes de la España contemporánea*. Madrid: Instituto de Estudios Políticos, 1959.

Aschmann, Birgit. *Preußens Ruhm und Deutschlands Ehre. Zum nationalen Ehrdiskurs im Vorfeld der preußisch-französischen Kriege des 19. Jahrhunderts*. München: Oldenbourg, 2013.

Bärsch, Georg. *Ferdinand v. Schill's Zug und Tod*. Leipzig: Brockhaus, 1860.

Beckett, Ian F.W. *Modern Insurgencies and Counter-Insurgencies: Guerrillas and their Opponents since 1750*. London: Routledge, 2001.

Bellver Amaré, Fernando. *Tomás de Zumalacárregui*. Madrid: Editorial Síntesis, 2010.

Benedict, Ruth. *The Chrysanthemum and the Sword*. 1st ed. 1946; London, Secker and Warburg, 1947.

Berlioz, Jacques and Marie-Anne Polo de Beaulieu, eds. *Les exempla médiévaux: Introduction à la recherche suivie des Tables critiques de l'Index exemplorum de F.C. Tubach*. Carcassonne: GARAE-Hésiode, 1992.

Berlioz, Jacques and Marie-Anne Polo de Beaulieu, eds. *Les exempla médiévaux: nouvelles perspectives*. Paris: H. Champion, 1998.

Binder von Krieglstein, [Carl]. *Ferdinand von Schill. Ein Lebensbild*. Berlin: Voss 1909.

Bock, Helmut. *Schills Rebellenzug 1809*. 4th ed. Berlin (East):1988.

Bois, Jean-Pierre. *Bugeaud*. Paris: Fayard, 1997.

Bois, Paul. *Les paysans de l'Ouest*. Paris and La Haye: Mouton, 1960.

Bolek, Francis (ed.). *Who's Who in Polish America*. 3rd ed. New York: Harbinger House, 1943.

Bonnet, Gabriel. *Les Guerres insurrectionnelles et révolutionnaires de l'Antiquité à nos jours*. Paris: Payot, 1958.

Boot, Max. *Invisible Armies: An Epic History of Guerrilla Warfare from Ancient Times to the Present*. New York: W.W. Norton & Co., 2013.

Borkenau, Franz. *The Spanish Cockpit*. Ann Arbor, MI: University of Michigan Press, 1963.

Broers, Michael. *Napoleon's Other War: Bandits, Rebels and their Pursuers in the Age of Revolutions*. Oxford: Peter Lang, 2010.

Burckhardt, Jacob. *La Civilisation en Italie au temps de la Renaissance*. Paris: Bartillat, 2012.

Burgo, Jaime del. *Historia de la primera guerra carlista: comentarios y acotaciones a un manuscrito de la época 1834–1839*. Pamplona: Institución Príncipe de Viana, 1981.

Cénat, Jean-Philippe. *Le roi stratège. Louis XIV et la direction de la guerre (1661–1715)*. Rennes: Presses universitaires de Rennes, 2010.

Cepeda Gómez, José. *El Ejército Español en la Política Española (1787–1843): conspiraciones y pronunciamientos en los comienzos de la España Liberal*. Madrid: Fundación Universitaria Española, 1990.

Chartier, R., M.-M. Compère, and D. Julia. *L'éducation en France du 16e au 18e siècle*. Paris: Sedes, 1976.

Clark, Christopher. *Iron Kingdom: The Rise and Downfall of Prussia, 1600–1947*. London: Allen Lane 2006, German translation *Preußen. Aufstieg und Niedergang, 1600–1947*. Munich: DVA, 2007.

Comellas García-Llera, José Luis. *El trienio constitucional*. Madrid: Ediciones Rialp, 1963.

Coverdale, John F. *The Basque Phase of Spain's First Carlist War*. Princeton, NJ: Princeton University Press, 1984.

Debrière, Édouard. *1793–1805 Projets et tentatives de débarquement dans les îles britanniques* [Projects and attempts to land in the British isles]. Vol. 1. Paris: Chapelot, 1901.

Delumeau, Jean. *La civilisation de la renaissance*. 1st ed. 1967; Paris: Arthaud, 1984.

Díaz Torrejón, Francisco. *Guerrilla, contra-guerrilla y delincuencia en la Andalucía napoleónica, 1810–1812*. Granada: Castillo, 2005.

Drévillon, Hervé. *L'impôt du sang. Le métier des armes sous Louis XIV*. Paris: Tallandrier, 2005.

Dodds, E.R. *The Greeks and the Irrational*. Berkeley: University of California press, 1951.

Duchhardt, Heinz. *Stein: eine Biographie*. Münster: Aschendorf 2007.

Duffy, Christopher. *The '45*. London: Orion/Phoenix, 2007.

Drévillon, Hervé. *Les rois absolus*. Paris: Belin, 2011.

Echternkamp, Jörg. *Der Aufstieg des deutschen Nationalismus (1770–1840)*. Frankfurt/M: Campus, 1998.

Elliott, Marianne. *Partners in Revolution: The United Irishmen and France*. New Haven, CT and London: Yale University Press, 1982.

Ellis, John. *A Short History of Guerrilla Warfare*. London: Ian Allan, 1975.

Esdaile, Charles. *The Peninsular War: A New History*. London: Penguin, 2003.

Esdaile, Charles J. *Fighting Napoleon: Guerrillas, Bandits and Adventurers in Spain 1808–1814*. New Haven, CT: Yale University Press, 2004.

Esdaile, Charles J. *Outpost of Empire: The Napoleonic Occupation of Andalucía, 1810–1812*. Norman, OK: University of Oklahoma, 2012.

Febvre, Lucien. *'Honneur et Patrie': Une enquête sur le sentiment d'honneur et l'attachement à la patrie*. Paris: Perrin, 1996.

Forrest, Alan. *Déserteurs et insoumis sous la Révolution et l'Empire*. Paris: Perrin, 1988.

Forrest, Alan. *The Legacy of the French Revolutionary Wars: The Nation-in-Arms in French Republican Memory*. Cambridge University Press, 2013.

Forrest, Alan. *Napoleon's Men: The Soldiers of the Revolution and Empire*. London and Hambledon: Continuum, 2002.

Forrest, Alan. *Soldiers of the French Revolution*. Durham, NC: Duke University Press, 1989.

Fraser, Ronald. *Napoleon's Cursed War: Popular Resistance in the Spanish Peninsular War*. London: Verso, 2007.

Frauenholz, Eugen von, ed. *Entwicklungsgeschichte des deutschen Heerwesens*. Munich: Beck, 1941.

Frémeaux, Jacaues. *De quoi fut fait l'empire: les guerres coloniales du XIXe siècle*. Paris: CNRS, collection Histoire, 2010.

Friederich, Rudolf. *Die Befreiungskriege 1813–1815*. Berlin: Mittler, 1911.

García Hernán, Enrique, Miguel Ángel de Bunes, Óscar Recio Morales, and Bernardo J. García García, eds. *Irlanda y la Monarquía hispánica: Kinsale 1601–2001: Guerra, Política, Exilio y Religión*. Madrid: Universidad de Alcalá, CSIS, 2002.

Gat, Azar. *A History of Military Thought: From the Enlightenment to the Cold War*. Oxford: Oxford University Press, 2001.

Gaucher, É. *La biographie chevaleresque: Typologie d'un genre (13e–15e siècle)*. Paris: Honoré Champion, 1994.

Girard, Alain. *Pourquoi la Vendée?* Paris: Armand Colin, 1990.

Godefroy, Frédéric. *Le Dictionnaire de l'ancienne langue française*. Paris: Wieweg et Bouillon, 1880–1902.

Gutiérrez Tellez, D. *Biografía de D. José Serrano Valdenebro, jefe de escuadra de la Real Armada Española, 1743–1814*. Ronda: Monte de Oca, 2008.

Guillon, Ernest. *La France et l'Irlande pendant la Révolution Hoche et Humbert* [France and Ireland during the Revolution Hoche and Humbert]. Paris: Colin, 1888.

Hagemann, Karen. *Männlicher Muth und teutsche Ehre. Nation, Militär und Geschlecht zur Zeit der Antinapoleonischen Kriege Preußens*. Paderborn: Ferdinand Schöningh, 2002.

Hahlweg, Werner. *Guerilla: Krieg ohne Fronten*. Stuttgart: Kohlhammer, 1968.

Hahlweg, Werner. *Preußische Reformzeit und revolutionärer Krieg*. Berlin: Mittler, 1962.

Halicz, Emanuel. *Partisan Warfare in 19th Century Poland: The Development of a Concept*. Translated from Polish by Jane Fraser. Odense: Odense University Press, 1975.

Heitzer, Heinz. *Insurrectionen zwischen Weser und Elbe: Volksbewaffnungen gegen die französische Fremdherrschaft im Königreich Westfalen 1806–1813*. Berlin: Rütten & Loening, 1959.

Heuser, Beatrice. *The Evolution of Strategy: Thinking War from Antiquity to the Present*. Cambridge: Cambridge University Press, 2010.

Heuser, Beatrice. *Rebellen, Partisanen, Guerrilleros: Asymmetrische Kriegführung von der Antike bis heute*. Paderborn: Schöningh, 2013.

Hobsbawm, Eric J. *The Age of Revolution: 1789–1848*. London: Abacus, 1962.

Hobsbawm, E.J. *Nations and nationalism since 1780*. Cambridge: Cambridge University Press, 1990.

Hobsbawm, Eric J. *Primitive Rebels*. Manchester: Manchester University Press, 1974.

Hoefer, M., ed. *Nouvelle Biographie générale*. Paris: Firmin-Diderot, 1856.

Höhn, Reinhard. *Revolution–Heer–Kriegsbild*. Darmstadt: Wittich, 1944.

Horrent, Jules. *La chanson de Roland dans les littératures française et espagnole au Moyen Âge*. Paris: Belles lettres, 1951.

Huguet, Edmond. *Dictionnaire de la langue française du seizième siècle*. Paris: Honoré Champion, reprint 1989.

Ibbeken, Rudolf. *Preußen 1807–1813. Staat und Volk als Idee und Wirklichkeit*. Cologne: Grote, 1970.

Ihl, Olivier. *Le Mérite et la République. Essai sur la société des émules*. Paris, Gallimard, 2007.

Iribarren, José María. *Espoz y Mina: El Liberal*. Madrid: Aguilar, 1967.

Israel, Jonathan. *A Revolution of the Mind: Radical Enlightenment and the Intellectual Origins of Modern Democracy*. Princeton, NJ: Princeton University Press, 2010.

James, Lawrence. *Warrior Race: A History of the British at War*. London: Abacus, 2001.

Jeanmougin, Bertrand. *Louis XIV à la conquête des pays-bas espagnols: La guerre oubliée, 1678–1684*. Paris: Economica, 2005.

Jessen, Olaf. *Preußens Napoleon? Ernst von Rüchel. Krieg im Zeitalter der Vernunft*. Paderborn: Schöningh, 2007.

Jouanna, Arlette. *Le devoir de révolte. La noblesse française et la gestation de l'État moderne, 1559–1661*. Paris: Fayard, 1989.

Jouanna, Arlette. *L'idée de race en France au 16e siècle et au début du 17e siècle (1498–1614)*. Lille: Atelier de Reproduction des thèses, 1976.

Jouanna, Arlette. *Ordre social: mythes et hiérarchies dans la France du 16e siècle*. Paris: Hachette, 1977.

Killy, Walter and Rudolf Vierhaus, eds. *Deutsche Biographische Enzyklopädie*. Munich: Saur, 1995.

Kunisch, Johannes. *Der kleine Krieg. Studien zum Heerwesen des Absolutismus*. Wiesbaden: Steiner, 1973.

Lefranc, Abel. *L'armée Française et la Renaissance. Un réformateur militaire au 16e siècle: Raymond de Fourquevaux*. Paris: Edouard Champion, 1916 [1st ed. in Revue du Seizième Siècle, t. III, 1915].

Lentz, Thierry. *Nouvelle histoire du Premier Empire*, vol 2, *L'effondrement du système napoléonien, 1810–1814*. Paris: Fayard, 2004.

Lubac, Henri. *Exégèse médiévale, les quatre sens de l'Écriture*. Paris: Ed. du Cerf, 1993.

Luh, Jürgen. *Der Große: Friedrich II. von Preußen*. Munich: Siedler, 2011.

Lynn, John. *Battle: A History of Combat and Culture*. Boulder CO: Westview, 2003.

Lynn, John. *Giant of the Grand Siecle: The French Army (1610–1715)*. Cambridge: Cambridge University Press, 1998.

Martín de Molina, Salvador. *Gaucín: 1742–1814: vida, trabajos, avatares, gesta y vindicación del Brigadier ... Don Josef Serrano Valdenebro, jefe de la guerrilla de la sierra meridional con cuartel general en la mencionada villa malagueña de Gaucín durante la guerra contra el invasor francés.* Córdoba: Cocoa, 1960–87.

Martin, Jean-Clément. *La Vendée de la mémoire, 1800–1980.* Paris: Seuil, 1989.

Martínez Laínez, Fernando. *Como lobos hambrientos. Los guerrilleros en la Guerra de la Independencia (1808–1813).* Madrid: Algaba, 2007.

Martinien, Aristide. *Tableaux par corps et par batailles des officiers tués et blesses pendant les guerres de l'empire, 1805–1815.* Paris: Librairie L. Fournier, 1908.

McCullough, Roy L. *Coercion, Conversion and Counterinsurgency in Louis XIV's France.* Leiden: Brill, 2007.

McMahon, Darrin. *Enemies of the Enlightenment: The French Counter-Enlightenment and the Making of Modernity.* New York: Oxford University Press, 2002.

Mattéi, Jean-Mathieu. *Histoire du droit de la guerre (1700–1819). Introduction à l'histoire du droit international.* Aix en Provence: Presses Universitaires d'Aix-Marseille, 2006.

Melgar, Francisco. *Pequeña historia de las guerras carlistas.* Pamplona: Editorial Gómez, 1958.

Möller, Horst. *Fürstenstaat oder Bürgernation: Deutschland 1763–1815.* Berlin: Siedler, 1989.

Mora-Lebrun, Francine. *L'Énéide médiévale et la naissance du roman.* Paris: PUF, 1994.

Moreti, Juan José. *Historia de L.M.N.Y.M.L Ciudad de Ronda.* Ronda: J.J. Moreti, 1867.

Nipperdey, Thomas. *Deutsche Geschichte 1800–1866. Bürgerwelt und starker Staat.* Munich: Beck, 1987.

Nitschke, Heinz G. *Die Preußischen Militärreformen 1807–1813.* Berlin (West): 1983.

Novak, Andrzej. *History and Geopolitics: A Contest for Eastern Europe.* Warsaw: Polish Institute of International Affairs, 2008.

Oyarzun, Román. *Historia del carlismo.* 3rd ed. Madrid: 1965.

Pérez, Garzón and Juan Sisino. *Milicia nacional y revolución burguesa: el prototipo madrileño, 1808–1874.* Madrid: Imprenta Sáez, 1978.

Pérez-Jean, Brigitte and Patricia Eichel-Lojkine, eds. *L'Allégorie, de l'Antiquité à la Renaissance.* Paris: Honoré Champion, 2004.

Pertz, Georg Heinrich, ed. *Das Leben des Feldmarschalls Grafen Neithardt von Gneisenau.* Berlin: Reimer, 1864.

Petitfrère, Claude. *Les bleus d'Anjou, 1789–92.* Paris: CTHS, 1985.

Petitfrère, Claude. *Les Vendéens d'Anjou, 1793.* Paris: Bibliothèque Nationale, 1981.

Peyrard, Christine. *Les Jacobins de l'Ouest: Sociabilité révolutionnaire et formes de politisation dans le Maine et la Basse-Normandie, 1789–1799.* Paris: Publications de la Sorbonne, 1996.

Picaud-Monnerat, Sandrine. *La Petite Guerre au XVIIIe Siècle.* Paris: Economica, 2010.

Pirala y Criado, Antonio. *Historia de la guerra civil y de los partidos liberal y carlista.* Vols. 6. Madrid: Turner, 1984.

Planert, Ute. *Der Mythos vom Befreiungskrieg. Frankreichs Kriege und der deutsche Süden. Alltag – Wahrnehmung – Deutung.* Paderborn: Schöningh, 2007.

Pocock, John G.A. *Le moment machiavélien. La pensée politique florentine et la tradition républicaine atlantique.* Paris: Presses Universitaires de France, 1997.

Poulain-Gautret, E. *La tradition littéraire d'Ogier le Danois après le 13e siècle, Permanence et renouvellement du genre épique medieval.* Paris: Honoré Champion, 2005.

Remírez de Esparza and Francisco Asín. *El carlismo aragonés, 1833–40.* Zaragoza: 1983.

Reynaud, Jean-Louis. *Contre-guerilla en Espagne (1808–1814): Suchet pacifie l'Aragon.* Paris: Economica, 1992.

Rink, Martin. *Vom Partheygänger zum Partisanen. Die Konzeption des kleinen Krieges in Preußen 1740–1813.* Frankfurt: Lang, 1999.

Risco, Alberto P. *Zumalacárregui en campaña. Según los documentos conservados por su secretario de estado mayor, don Antonio Zaratiegui*. Madrid: Imprenta de José Murillo, 1935.

Rorive, Jean-Pierre. *Les misères de la guerre sous le Roi-Soleil: les populations de Huy, de Hesbaye et du Condroz dans la tourmente du Siècle de malheur*. Liège: Les Éditions de l'Université de Liège, 2000.

Rößler, Helmut. *Graf Stadion*. Vienna, Munich: 1966.

Santirso, Manuel, ed. *Joseph Tański: el informe Tański y la guerra civil carlista de 1833–1840*. Madrid: Ministerio de Defensa, 2011.

Satterfield, George. *Princes, Posts and Partisans: The Army of Louis XIV and Partisan Warfare in the Netherlands (1673–1678)*. Leyden: Brill, 2003.

Schmitt, Carl. *Theory of the Partisan: intermediate commentary on the concept of the political*. Translated by G.L. Ulmen. New York: Telos, 2007.

Schulze, Hagen. *Staat und Nation in der europäischen Geschichte*. 2nd ed. Munich: Beck, 1995.

Secher, Reynald. *Le génocide franco-français: La Vendée-Vengé*. Paris: Presses Universitaires de France, 1986.

Siebert, Karl. *Hanauer Biographien aus drei Jahrhunderten*. Hanau: Hanauer Geschichtsverein, 1919.

Skinner, Quentin. *The Foundations of Modern Political Thought*. Cambridge: Cambridge University Press, 1978. Translated into French as *Les Fondements de la pensée politique modern*. Paris: Albin Michel, 2001.

Smith, Jay M. *Monsters of the Gévaudan: The Making of a Beast*. Cambridge, MA: Harvard University Press, 2011.

Stamm-Kuhlmann, Thomas. *König in Preußens großer Zeit. Friedrich Wilhelm III. der Melancholiker auf dem Thron*. Berlin: Siedler, 1992.

Starkey, Armstrong. *European and Native American warfare, 1675–1815*. London: University College London Press, 1998.

Strubel, Armand. *'Grant senefiance a': Allégorie et littérature au Moyen Âge*. Paris: Champion, 2002.

Stuart, Jones and Edwyn Henry. *The Last Invasion of Britain*. Cardiff: University of Wales Press, 1950.

Szabó, János B. and Ferenc Tóth. *Mohács (1526). Soliman le Magnifique prend pied en Europe central*. Paris: Economica, 2009.

Tallett, Frank. *War and Society in Early Modern Europe, 1495–1715*. 1st ed. 1992; London: Routledge, 2001.

Thiele, Gerhard, ed. Gneisenau. *Leben und Werk des königlich-preußischen General-feldmarschalls. Eine Chronik*. Potsdam: Verlag für Berlin-Brandenburg, 1999.

Tilly, Charles. *The Vendée*. London: Arnold, 1964.

Verrier, Frédérique. *Les armes de Minerve. L'humanisme militaire dans l'Italie du Nord du 16ᵉ siècle*. Paris: Presses Universitaires de Paris-Sorbonne, 1997.

Walter, Dierk. *Preußische Heeresreformen 1807–1870. Militärische Innovation und der Mythos der Roonschen Reform*. Paderborn: Ferdinand Schöningh, 2003.

Walzer, Michael. *Just and Unjust Wars: A Moral Argument with Historical Illustrations*. New York: Basic Books, 1977.

Wehler, Hans-Ulrich. *Deutsche Gesellschaftsgeschichte*. Vol. 1, 2nd ed. Munich: Beck, 1989.

Weighley, Russell F. *The American Way of War: A History of United States Military Strategy and Policy*. Bloomington, IN: Indiana University Press, 1973.

Wolff, Étienne, ed. *Fulgence le mythographe: Virgile dévoilé*. Villeneuve-d'Ascq: Presses Universitaires du Septentrion, 2009.

Wohlfeil, Rainer. *Spanien und die deutsche Erhebung 1808–1814*. Wiesbaden: Franz Steiner Verlag, 1965.
Zimmermann, Paul. *Der Schwarze Herzog Friedrich Wilhelm von Braunschweig*. Hildesheim: Lax, 1936.

Edited volumes

Chanet, Jean-François and Christian Windler, eds. *Les ressources des faibles. Neutralités, sauvegardes, accommodements: micro-histoire des arrangements face à la guerre et à l'occupation*. Rennes: Presses Universitaires de Rennes, 2009.
Coutau-Bégarie, Hervé, ed. *Stratégies irrégulières*. Paris: Économica, 2010.
Coutau-Bégarie, Hervé and Charles Doré-Graslin, eds. *Histoire militaire des Guerres de Vendée*. Paris: Economica, 2010.
Esdaile, Charles, ed. *Popular Resistance in the French Wars: Patriots, Partisans and Land Pirates*. Basingstoke: Palgrave Macmillan, 2005.
Gérard, Alain and Thierry Heckmann, eds. *La Vendée dans l'histoire*. Paris: Perrin, 1994.
Haythornthwaite, Phillip, ed. *In the Peninsula with a French Hussar: Memoirs of the War of the French in Spain*. London: Greenhill Books, 1990.
Priesdorff, Kurt von, ed. *Soldatisches Führertum*. Hamburg: Hanseatische Verlagsanstalt, 1938.
Tallett, Frank and D.J.B. Trim, eds. *European Warfare, 1350–1750*. Cambridge: Cambridge University Press, 2010.

Journal articles and chapters in edited volumes

Ahlstrom, John D. 'Captain and Chef de brigade William Tate; South Carolina Adventurer'. *South Carolina Historical Magazine* 88, no. 4 (1987): 183–91.
Antoche, Emmanuel-Constantin. 'Les guerres irrégulières dans les principautés de Moldavie et de Valachie (XIVe-XVe siècle)'. In *Stratégies irrégulières*, ed. Hervé Coutau-Bègarie, 160–84. Paris: Économica, 2010.
Aróstegui, Julio. 'La aparición del carlismo y los antecedents de la guerra' in *Historia de España: La era isabelina y el sexenio democrático. 1834–1874*, Vol. XXXIV, essays for Ramón Menéndez Pidal, edited by José María Jover Zamora. Madrid: Espasa-Calpe, 1981.
Artola-Gallego, Miguel. 'La guerra de guerrillas.' In *La Guerra de la Independencia en el mosaico peninsular 1808–1814*, edited by Cristina Borreguero Beltrán, 357–66, Burgos, 2010.
Bartlett, Thomas. 'Defence, Counter-Insurgency and Rebellion: Ireland, 1793–1803'. In *A Military History of Ireland*, edited by Thomas Bartlett and Keith Jeffery, 247–490. Cambridge: Cambridge University Press, 1996.
Berlioz, Jacques and Marie-Anne Polo de Beaulieu. 'Les prologues des recueils d'exempla'. In *Les prologues médiévaux*, edited by Jacqueline Hamesse, 275–320. Turnhout: Brepols, 2000.
Blanchard, Joël. 'Écrire la guerre au 15e siècle'. *Le moy âge en français* 24–25 (1989): 7–21.
Boycott-Brown, Martin. 'Guerrilla Warfare avant la lettre: Northern Italy, 1792–97'. In *Popular Resistance in the French Wars: Patriots, Partisans and Land Pirates*, edited by Charles Esdaile, 45–66. London: Palgrave Macmillan, 2004.
Cénat, Jean-Philippe. 'Le Ravage du Palatinat: politique de destruction, stratée de cabinet et propaganda au début de la guerre de la Ligue d'Augsbourg'. *Revue historique* 633 (2005): 97–132.
Clark, Anna. 'The Rhetoric of Masculine Citizenship'. In *Representing Masculinity: Male Citizenship in Modern Western Culture*, edited by Stefan Dudink, Karen Hagemann, and Anna Clark, 4–16. New York: Palgrave Macmillan, 2007.

Come, Donald R. 'French Threats to British Shores, 1793–1798'. *Military Affairs* 16, no. 4 (1952): 174–88.

Contamine, Philippe. 'C'est un très périlleux héritage que guerre'. *Vingtième siècle*. Revue d'histoire 3 (1984): 5–15.

Coutansais, F. 'La guerre des Géants vue par les Bleus'. *Revue du Bas-Poitou* 74 (1963): 427–37.

Coutau-Bégarie, Hervé. 'Guerres irrégulières: de quoi parle-t-on?' In *Stratégies Irrégulières*, edited by Hervé Coutau-Bégarie, 13–47. Paris: Economica, 2010.

Czigány, István. 'Tradition et modernité dans les affaires militaires du royaume de Hongrie'. In *Stratégies irrégulières*, edited by Hervé Coutau-Bégarie, 277–278. Paris: Économica, 2010.

Deruelle, Benjamin. 'Enjeux politiques et sociaux de la culture chevaleresque au 16e siècle: les prologues de chansons de geste imprimées'. *Revue Historique* 655 (2010): 551–576.

Deruelle, Benjamin. 'Des limites imperceptibles à l'exercice de la force au 16e siècle: théorie et pratiques de la "bonne guerre" dans les armées du roi de France'. In *Faire la guerre, faire la paix*, Paris: CTHS, 2013, http://cths.fr/co/communication.php?id=5420

Deruelle, Benjamin. '"Pour Dieu, le roi et l'honneur": Ethos chevaleresque, mérite et récompense au 16e siècle'. *Hypothèses* 13 (2008): 209–220.

Diego García, Emilio de. 'Balance de un conflicto marcado por la complejidad'. In *Andalucía en Guerra, 1808–1814*, edited by José Miguel Delgado Barrado and María Amparo López Arandia. Jaén: Universidad de Jaén, 2010.

Drévillon, Hervé. 'L'âme est à Dieu et l'honneur à nous'. *Revue historique* 654 (2010): 361–395.

Drévillon, Hervé. 'Courtilz de Sandras et les valeurs militaires de la noblesse à la fin du règne de Louis 14'. In *Combattre, gouverner, écrire*, edited by Yves-Marie Bercé, Philippe Contamine, André Corvisier, 351–67. Paris: Commission Française d'Histoire Militaire, Institut de Stratégie Comparée Paris I Sorbonne & Ed. Économica, 2003.

Dupuy, Roger. 'Ignorance, fanatisme, et contre-révolution'. In *Les résistances à la Révolution*, edited by François Lebrun and Roger Dupuy, 35–40. Paris: Imago, 1987.

Esdaile, Charles J. 'War and Politics in Spain, 1808–1814'. *The Historical Journal* 31, no. 2 (1988): 295–317.

Forrest, Alan. 'La guerre de l'Ouest vue par les soldats républicains'. In *La guerre civile entre Histoire et Mémoire*, edited by Jean-Clément Martin, 91–99. Nantes: Ouest éd, 1995.

Forrest, Alan. 'The Ubiquitous Brigand: The Politics and Language of Repression'. In *Popular Resistance in the French Wars: Patriots, Partisans and Land Pirates*, edited by Charles Esdaile, 25–43. Basingstoke: Palgrave Macmillan, 2005.

Fratani, Dominique. 'Les chevaux des Gonzagues à la bataille de Fornoue'. In *Le cheval et la guerre*, edited by Daniel Roche and Daniel Reytier, 45–53. Paris: Association pour l'académie d'art équestre de Versailles.

Gastey [Paul], General. 'L'étonnante aventure de l'Armée d'Irlande 1798'. *Revue Historique des Armées* 4 (1952): 19–32.

Hargreaves, Andrew L. 'The Advent, Evolution and Value of British Specialist Formations in the Desert War, 1940–1943'. *Global War Studies* 7, no. 2 (2010): 7–61.

Hamesse, Jacqueline, ed. 'Les exempla médiévaux'. In *Introduction à la recherche suivie des Tables critiques de l'Index exemplorum de F.C. Tubach*. Carcassonne: GARAE-Hésiode, 1992.

Heuser, Beatrice. 'Covert Action within British and American Concepts of "Containment'.
In *British Intelligence, Strategy and the Cold War, 1945–51*, edited by Richard Aldrich, 65–84. London: Routledge, 1992.

Heuser, Beatrice. 'Guibert (1744–1790): Prophet of Total War?' In *War in an Age of Revolution: The Wars of American Independence and French Revolution, 1775–1815*, edited by Stig Förster and Roger Chickering, 49–67, Cambridge University Press, 2010.

Heuser, Beatrice. 'Small Wars in the Age of Clausewitz: Watershed between Partisan War and People's War'. *Journal of Strategic Studies* 33, no. 1 (2010): 137–60.

Heuser, Beatrice. 'Victory, Peace, and Justice: The Neglected Trinity'. *Joint Forces Quarterly* 69 (2013): 6–12.

Hubrecht, Georges. 'La guerre juste dans la doctrine chrétienne, des origines au milieu du 16e siècle'. In *Recueils de la Société Jean Bodin pour l'histoire comparative des institutions*, 15, 107–123, 'La paix deuxième partie', 2nd ed. Paris: Dessain et Tolra, 1984.

Joes, Anthony James. 'Insurgency and Genocide: La Vendée'. *Small Wars & Insurgencies* 9, no. 3 (1998): 17–45.

Kaempff, Sebastian. 'Lost through Non-Translation: Bringing Clausewitz's Writings on "New Wars" Back In'. *Small Wars and Insurgencies* 22, no. 4 (2011): 548–73.

Keller, H.-Er. 'Autour de Galien le Restoré'. In *De l'aventure épique à l'aventure Romanesque*, edited by Jacques Chocheyras, 77–84. Bern: Lang, 1997.

Kleinman, Sylvie. '"Un brave de plus": La carrière militaire de Theobald Wolfe Tone, héros du nationalisme irlandais et officier français, 1796–1798', *Revue Historiques des Armées* 253 (2008): 55–65.

Kleinman, Sylvie. '"Amidst Clamour and Confusion": Civilian and Military Linguists at War in the Franco-Irish Campaigns against Britain (1792–1804)'. In *Languages and the Military Alliance, Occupation and Peace Building*, edited by Hilary Footitt and Michael Kelly, 25–46. Basingstoke: Palgrave Macmillan, 2012.

Kleinman, Sylvie. 'Libérer ou exploiter? L'Irlande dans la stratégie diplomatique et militaire de la France, 1792–1804'. In *Les Horizons de la politique extérieure française Périphéries et espaces seconds XVIe – XXe siècles*, edited by Frédéric Dessberg and Eric Schnackenbourg, 283–96. Berne: Peter Lang, 2011.

Körner, Theodor. 'Männer und Buben'. In *Leyer und Schwert*, by Theodor Körner, 77–81. Wien, 1814. http://archive.org/stream/leyerundschwerdt00kr

Koselleck, Reinhart. 'Einleitung'. In *Geschichtliche Grundbegriffe*, Vol. 1, edited by Otto Brunner, Werner Conze, and Reinhart Koselleck, XIII–XXVII. Stuttgart: Klett-Cotta, 1972.

Koyré, Alexandre. 'Galilée et Platon'. *Études d'histoire de la pensée scientifique*. Paris: Gallimard, 1985 (1st ed. 1973), 166–95.

Lacroix-Leclerc, Jérôme and Éric Ouellet. 'The Petite Guerre in New France 1660–1759: An Institutional Analysis'. *Canadian Military Journal* 11, no. 4 (2011): 48–54.

Langendorf, Jean-Jacques. 'Landwehr et Landsturm. Une armée d'ombres et une armee à l'ombre de l'armee'. In *Stratégies Irrégulières*, edited by Hervé Coutau-Bégarie, 388–405. Paris: Economica, 2010.

Lawrence, Mark. 'Las viudas de Comares: un caso de radicalismo popular en la Málaga Liberal'. In *Visiones del Liberalismo: política, identidad y cultura en la España del siglo XIX*, edited by Alda Blanco and Guy Thomson. Valencia: Editorial Acribia, 2008.

Lawrence, Mark. 'Peninsularity and Patriotism: Spanish and British Approaches to the Peninsular War, 1808–14'. *Historical Research* 85 (2012): 453–68.

Lepetit, Gildas. 'Soumettre les arrières de l'armée. L'action de la Gendarmerie impériale dans la pacification des provinces septentrionales de l'Espagne (1809–1814)'. In *Stratégies Irrégulières*, edited by Hervé Coutau-Bégarie, 372–87. Paris: Economica, 2010.

Lynn, John. 'How War Fed War: The Tax of Violence and Contributions during the Grand Siecle'. *The Journal of Modern History* 65 (1993): 286–310.

Mantovani, Mauro. 'Der 'Volksaufstand': Vorstellungen und Vorbereitungen der Schweiz im 19. und 20. Jahrhundert'. *Military Power Revue der Schweizer Armee* 1 (2012): 52–60.

Martín de Molina, Salvador. 'La guerrilla en la Serranía vista por cinco de sus protagonistas'. *Cuadernos del Bicentenario* 10 (2010): 191–207.

Martin, Jean-Clément. 'Introduction'. In *La guerre civile entre Histoire et Mémoire*, edited by Jean-Clément Martin, 15–20. Nantes: Ouest Editions, 1995.

Martin, Jean-Clément. 'Vendée, guerre de'. In *Dictionnaire de la Contre-révolution*, edited by Jean-Clément Martin, 500–505. Paris: Perrin, 2011.

Mitchell, Harvey. 'Resistance to the Revolution in Western France'. *Past and Present* 63 (1974): 94–131.

Mönch, Winfried. 'Rokokostrategen. Ihr negativer Nachruhm in der Militärgeschichtss-chreibung des 20. Jahrhunderts'. In *Die Kriegskunst im Lichte der Vernunft. Militär und Aufklärung im 18. Jahrhundert* edited by Daniel Hohrath & Klaus Gerteis. Aufklärung 11, no. 2 (1999): 76–98.

Muraise, Eric. 'L'insurrection royaliste de l'Ouest, 1791–1800'. In *Histoire militaire des Guerres de Vendée,* edited by Hervé Coutau-Bégarie and Charles Doré-Graslin. Paris: Economica, 2010.

O'Connell, J. J. 'Ireland in Strategical Geography'. *Irish Quarterly Review* 26, no. 104 (1937): 595–608.

Pagès, G. 'Lettres de requis et volontaires de Coutras en Vendée et en Bretagne'. *Revue historique et archéologique du Libournais* 190 (1983): 153–62.

Paquette, J.-M. 'Introduction'. In *Typologie des sources du Moyen Age occidental*. Vol. 49. L'épopée, edited by R. Boyer et al., Turnhout: Brepols, 1988.

Paret, Peter. 'Colonial Experience and European Military Reform at the End of the Eighteenth Century'. Originally 1964, republished in *Historical Research* 37, no. 95 (2007): 47–59.

Paret, Peter. 'The Relationship between the Revolutionary War and European military thought and practice in the second half of the 18th century'. In *Reconsiderations on the Revolutionary War*, edited by Don Higginbotham, 144–157, 208–210. Westport, CT: Greenwood Press, 1978.

Paris, G. 'Galien'. In *Histoire littéraire de la France*, Paris: Imprimerie nationale, 1881, t. 28, pp. 221–39.

Pepper, Simon. 'Aspects of operational Art: Communications, Cannon, and Small War'. In *European Warfare, 1350–1750*, edited by Frank Tallett and D. J. B. Trim, 181–202. Cambridge: Cambridge University Press, 2010.

Peschot, Bernard. 'Les "lettres de feu": la petite guerre et les contributions paysannes au XVIIe siècle'. In *Les villageois face à la guerre, XIVe–XVIIIe siècle*, edited by Christian Desplat, 129–42. Toulouse: Presses Universitaires du Mirail, 2002.

Peschot, Bernard. 'La petite guerre au 16e siècle: formes, style et contacts dans l'Occident méditerranéen'. In *Les armes et la toge*, Montpellier: CHMEDN, 1997.

Peschot, Bernard. 'La Guérila à l'époque modern'. *Revue Historique des Armées* 210 (1998).

Peschot, Bernard. 'La Notion de petite guerre en France (18e siècle)'. *Les cahiers de Montpellier* 28 (1993-II).

Petitfrère, Claude. 'The origins of the civil war in the Vendée'. *French History* 2 (1988): 187–207.

Picaud-Monnerat, Sandrine. 'La "guerre de partis" au 17e siècle en Europe'. *Stratégique* 88 (2007): 101–46.

Picaud-Monnerat, Sandrine. 'La "guerre de partis" au XVIIe siecle en Europe'. In *Stratégies irrégulières*, edited by Hérve Coutau-Bégarie, 202–34. Paris: Economica, 2010.

Picaud-Monerat, Sandrine. 'Partisan warfare, "war in detachment": la "petite guerre" vue d'Angleterre (18e siècle)'. *Stratégique* 84 (2004): 13–59.

Pionchon, Pauline. 'La Généalogie des dieux païens entre le Décaméron et les nouvelles des humanistes du premier 15e siècle'. *Cahiers d'études italiennes* 10 (2010): 55–78.

Pitt-Rivers, Julian. 'La maladie de l'honneur'. In *L'honneur. Image de soi ou don de soi: une image equivoque*, edited by Marie Gautheron, Paris: Autrement, 1991.

Ravaille, Luc. 'La petite guerre dans les commentaires de Monluc'. In *Mediterrán tanulmányok Études sur la région méditerranéenne*, Vol. XIV. Szeged: 2005.

Rink, Martin. 'Ein Patriot und Partisan. Ferdinand von Schill als Freikorpskämpfer neuen Typs'. In *Für die Freiheit–gegen Napoleon, Ferdinand von Schill, Preußen und die deutsche Nation edited by Veit Veltzke*, 65–106. Cologne: Böhlau, 2009.

Rink, Martin. 'Preußisch-deutsche Konzeptionen zum Volkskrieg im Zeitalter Napoleons'. In *Reform – Reorganisation – Transformation. Zum Wandel in deutschen Streitkräften von den preußischen Heeresreformen bis zur Transformation der Bundeswehr*, edited by Karl-Heinz Lutz, Martin Rink and Marcus von Salisch, 65–87. Munich: Oldenbourg, 2010.

Rink, Martin. 'Graf Wilhelm von Schaumburg-Lippe. Ein sonderbarer Duodezfürst als militärischer Innovator'. In *Die Schlacht bei Minden. Weltpolitik und Lokalgeschichte*, edited by Martin Steffen, 137–155, 237–243. Minden: J.C.C. Bruns, 2009.

Rink, Martin. 'Der kleine Krieg: Entwicklungen und Trends asymmetrischer Gewalt 1740 bis 1815'. *Militärgeschichtliche Zeitschrift* 65 (2006): 355–88.

Rink, Martin. 'The Partisan's Metamorphosis from Freelance Military Entrepreneur to German Freedom Fighter, 1740–1815'. *War in History* 17 (2010): 6–36.

Rose, Richard. 'The French at Fishguard: Fact, Fiction, Folklore'. *Transactions of the Honourable Society of Cymmrodorion* new series 9 (2003): 74–105.

Roura i Aulinas, Lluís. 'Jacobinos y jacobinismo en los primeros momentos de la revolución liberal española'. In *Revolución y democracia: el jacobinismo europeo*, edited by Lluís Roura i Aulinas and Irene Castells, 55–84. Madrid: Ediciones del Orto, 1995.

Rowe, Michael. 'Civilians and warfare during the French Revolutionary Wars'. In *Daily Lives of Civilians in Wartime Europe, 1618–1900*, edited by Linda S. Frey and Marsha L. Frey, 110–120. Westport, CT: Greenwood, 2007.

Sant Cassia, Paul. 'Banditry, Myth and Terror in Cyprus and Other Mediterranean Societies'. *Comparative Studies in Society and History* 35, no. 4 (1993): 773–95.

Sarmant, Thierry. 'Une seconde cavalerie: les dragons de Louis XIV et de Louvois'. In *Le cheval et la guerre du XVe au XX siècle*, edited by Daniel Roche, 232–41. Paris: Association pour l'académie d'art équestre de Versailles, 2002.

Scheipers, Sibylle. 'The Status and Protection of Prisoners of War and Detainees'. In *The Changing Character of War*, edited by Hew Strachan and Sibylle Scheipers, 394–409. Oxford: Oxford University Press, 2011.

Schmidt, Peer. 'Der Guerrillero: die Entstehung des Partisanen in der Sattelzeit der Moderne – eine atlantische Perspektive, 1776–1848'. *Geschichte und Gesellschaft* 29 (2003): 161–90.

Selig, Robert A. and David Curtis Skaggs. 'Introductory Essay'. to Ewald, Johann. *Treatise on Partisan Warfare*. New York, NY: Greenwood Press, 1991, 1–38.

Sikora, Michael. 'Scharnhorst. Lehrer, Stabsoffizier, Reformer'. In *Reform – Reorganisation – Transformation. Zum Wandel in deutschen Streitkräften von den preußischen Heeresreformen bis zur Transformation der Bundeswehr*, edited by Karl-Heinz Lutz, Martin Rink, and Marcus von Salisch, 43–64. Munich: Oldenbourg, 2010.

Storrs, Christopher. S. 'The Military Revolution and the European Nobility, c. 1600–1800'. In *War in History*, 3 (1996).

Thimme, Friedrich. 'Zu den Erhebungsplänen der preußischen Patrioten im Sommer 1808'. *Historische Zeitschrift* 86 (1901): 79–110.

Tilly, Charles. 'The Analysis of a Counter-Revolution'. *History and Theory* 3 (1963): 30–58.

Tilly, Charles. 'Local Conflicts in the Vendée before the Rebellion of 1793'. *French Historical Studies* 2 (1961): 209–31.

Tóth, Ferenc. 'Régularité et irrégularité dans la guerre d'indépendance hongroise au début du XVIIIe siècle'. In *Stratégies Irrégulières*, edited by Hervé Coutau-Bégarie, 279–92. Paris: Economica, 2010.

Vauchez, André. 'La notion de guerre juste au Moyen Âge'. In *Dissuasion nucléaire et conscience chrétienne*, 19, 9–22, Paris: Beauchesne, 1984.

Veltzke, Veit. 'Zwischen König und Vaterland. Schill und seine Truppen im Netzwerk der Konspiration'. In *Für die Freiheit–gegen Napoleon. Ferdinand von Schill, Preußen und die deutsche Nation*, edited by Veit Veltzke, 107–55. Cologne: Böhlau, 2009.

Viguerie, Jean de. 'La Vendée et les Lumières: les origines intellectuelles de l'extermination'. In *La Vendée dans l'histoire,* edited by Alain Gérard et Thierry Heckmann, Paris: Perrin, 1994, 36–43.

Wahnich, Sophie. 'La logique de l'exclusion révolutionnaire'. In *La guerre civile entre Histoire et Mémoire*, ed. Jean-Clément Martin, Nantes: Ouest éd., 1995.

Weber, Hermann. 'La stratégie de la terre brûlée: le cas du Palatinat en 1689'. In *La Vendée dans l'histoire*, edited by Alain Gérard and Thierry Heckmann, 190–200. Paris: Perrin, 1994.

Unpublished theses

Beresford de Paor, Marcus. 'Ireland in French strategy 1692–1789'. MLitt diss. University of Dublin, 1975.

Deruelle, Benjamin. 'De papier, de fer et de sang: chevaliers et chevalerie à l'épreuve du 16e siècle (ca. 1460 – ca. 1620)'. PhD diss. University Paris 1 Panthéon-Sorbonne, 2011. [Publication forthcoming as *De papier, de fer et de sang*. Paris: Presses universitaires de la Sorbonne, expected 2015.]

Fonck, Bertrand. 'Le maréchal de Luxembourg (1628–1695) et le commandement des armées: carrière des armes et pratique de la guerre sous Louis XIV'. PhD diss. Paris-Sorbonne, 2011. [Publication forthcoming as *Le maréchal de Luxembourg*. Champ Vallon: Seyssel, 2014.]

Guinier, Arnaud. 'L'Honneur du soldat. La discipline militaire en débat dans la France des Lumières (ca. 1748 – ca. 1789)'. 2 vols. PhD diss. University of Poitiers, 2012.

Holeindre, Jean-Vincent. 'Le renard et le lion. La ruse et la force dans le discours de la guerre'. PhD diss., École des Hautes Études en Sciences Sociales, 2010.

Kleinman, Sylvie. 'Translation the French Language and the United Irishmen 1792–1804'. PhD diss. Dublin City University, 2005.

Peschot, Bernard. 'La guerre buissonnière: partis et partisans dans la petite guerre (XVIe–XVIIIe siècle)'. PhD diss., University of Montpellier III, 1999.

Tyson, Peter J. 'The role of Republican and patriote discourse in the insurrection of the Vendée'. Ph.D. University of York, 1994.

Index

INDEX

Van der Meulen 46
Varca, Martín 117–18
Veas, Antonio García de 86
Vegetian tradition 21–2
Vendée 64–77, 178–85; influence of 178–85
vengeance 60, 75
Versailles 45
via Ireland to Britain 50–52
Ville, Antoine de la 8, 19
Virée de Galerne 65–6
Volkskriegsee 'people's war'
volunteers 52, 113, 146–7
Voss, Julius von 133

Wahnich, Sophie 73
Wall, Patrick 51
'war in absolute perfection' 174
War of the Austrian Succession 93–4
War of Reunions (1683–1684) 38
warfare revolutions 174–5

wars of revolution 92–5, 104
Washington, George 132–3, 178, 180
watershed between partisan and people's
   wars 122–45
way of 'people's war' 104
Wees, Hans van 159
West Point 178
Westermann, François-Joseph 178–9
White Terror 163–4
Wiriyamu massacre 165, 167–8
Wounded Knee 165

xenophobia 11

Yorck, Hans David Ludwig von 94–5, 101

Zaldívar, Pedro 82, 88
Zumalacárregui, Tomás de 111–13, 116
Zurbano Baras, Martín *see*Varca, Martín
Zweihänder 7